WHAT'S HAPPENING

in the

Mathematical Sciences

Volume

DANA MACKENZIE
BARRY CIPRA

WHAT'S HAPPENING

in the

Mathematical Sciences

AMERICAN MATHEMATICAL SOCIETY
www.ams.org

2000 *Mathematics Subject Classification*:
00A06

ISBN-13: 978-0-8218-3585-2
ISBN-10: 0-8218-3585-8

Cover design by Erin Murphy of the
American Mathematical Society.

The cover and the frontmatter for this publication were prepared using the Adobe®
CS2® suite of software. The articles were
prepared using TEX. TEX is a trademark of
the American Mathematical Society.

About the Authors

DANA MACKENZIE is a freelance mathematics and science
writer who lives in Santa Cruz, California. He received his
Ph.D. in mathematics from Princeton University in 1983 and
taught for thirteen years at Duke University and Kenyon
College before he decided to change careers and become a
writer. He finished the Science Communication Program at
the University of California at Santa Cruz in 1997. Since then
he has written on a semi-regular basis for such magazines as
Science, New Scientist, SIAM News, Discover, and *Smithsonian*.
He is also a contributing editor for *American Scientist*. His first
book, *The Big Splat, or How Our Moon Came to Be*, was published by John Wiley & Sons and named to *Booklist* magazine's
Editors' Choice list for 2003.

BARRY CIPRA, who also did the writing for volumes 1–5 of
What's Happening in the Mathematical Sciences, is a freelance
mathematics writer based in Northfield, Minnesota. He is
currently a Contributing Correspondent for *Science* magazine
and also writes regularly for *SIAM News*, the newsletter of the
Society for Industrial and Applied Mathematics. He received
the 1991 Merten M. Hasse Prize from the Mathematical
Association of America for an expository article on the Ising
model, published in the December 1987 issue of the *American
Mathematical Monthly*. His book, *Misteaks...and how to find
them before the teacher does...* (a calculus supplement), is
published by AK Peters, Ltd.

Contents

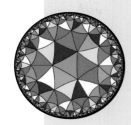

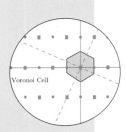

Voronoi Cell

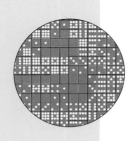

Introduction

MATHEMATICS IS THE MOST WIDELY STUDIED SUBJECT in the world. It is a subject with a distinguished and ancient past, but also a subject with an active present, affecting nearly every aspect of modern life.

Yet in spite of this, most people are skeptical when told that new mathematics is being created today. This is easy to understand when one realizes that the general public knows mathematics through arithmetic, high school algebra, and (perhaps) calculus. The most recent of these topics is 350 years old.

Volume 6 of the series *What's Happening in the Mathematical Sciences* provides evidence of the active nature of mathematics by highlighting ten notable research topics from the past few years.

The articles in this volume also highlight some common themes that run throughout the history of mathematical research. One theme is the surprising unity of mathematics—that solving established problems in one branch of mathematics may require tools and insights from what were thought to be unrelated branches of mathematics ("Millennium Problems," p. 2 and "New Insights," p. 52). Another is its unreasonable effectiveness—that abstract mathematics applied to physical systems often yields deep understanding that is sometimes hard to explain ("Navier-Stokes Equations," p. 78, "Mysteries of Insect Motion," p. 86, and "Brownian Motion," p. 100). And yet another theme is the hidden insight of mathematicians—that solving old problems often requires new insights and novel approaches that originate with young as well as established mathematicians, often through their joint efforts ("Venn-erable Problem," p. 40, "New Insights," p. 52, "Quadratic Number Fields and Beyond," p. 66, and "Classifying Hyperbolic Manifolds," p. 14).

This is truly current research. Three of these articles ("Millennium Problems," p. 2, "New Insights," p. 52, and "Brownian Motion," p. 100) describe work that was recognized with Fields Medals at the 2006 International Congress of Mathematicians held in August 2006 in Madrid, Spain. Fields Medals, awarded only every four years, are the analogue in mathematics of a Nobel Prize, but are awarded only to mathematicians under the age of 40.

This is a book for those who think of mathematics as a dull and dead subject. It will convince you that mathematics is both fascinating and alive, touching many parts of your everyday life and promising to grow even more lively in the future.

John Ewing
Publisher, American Mathematical Society

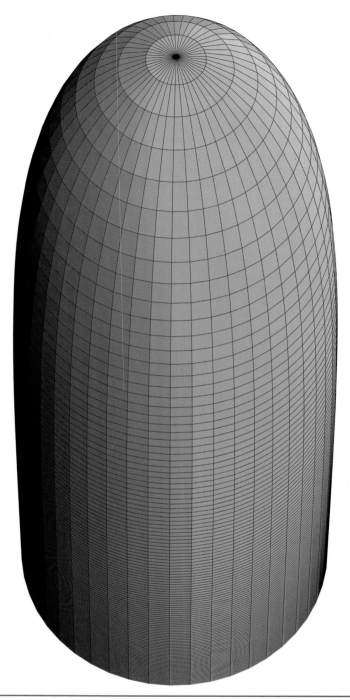

The "cigar." *This is a solution of the Ricci flow equation that remains stationary for all time, and therefore does not form a narrow tendril that can be broken off by surgery. One of the first key steps of Perelman's proof was to show that the cigar does not arise as a limit of Ricci flow on three-dimensional manifolds. (Graphic created by Michael Trott using Mathematica.)*

First of Seven Millennium Problems Nears Completion

Barry Cipra

IN THE MILLENNIAL YEAR OF 2000, the Clay Mathematics Institute focused the world's attention on seven mathematical problems of exceptional historical and practical interest, by offering a million-dollar prize for the solver of any of them. (See "Think and Grow Rich," *What's Happening in the Mathematical Sciences*, Volume 5). The seven problems were designated Millennium Prize Problems. Within three years, the first serious contender for one of the million-dollar prizes emerged. It now appears likely that the list of Millennium Problems will shrink from seven to six before the end of the millennium's first decade.

In November 2002 and March 2003, Grigory Perelman of the Steklov Institute in Moscow posted a pair of papers that outlined the key steps in settling a century-old topological problem known as the Poincaré conjecture. Because the two papers bring together ideas from disparate fields, and leave many details for the reader to fill in, experts have found them to be tough going. However, in more than three years of scrutiny, none of them have found any gaps that seriously jeopardized Perelman's claim to proving the theorem. Several experts seem to be cautiously edging towards pronouncing the proof complete.

Perhaps just as importantly, Perelman has introduced a raft of new techniques in differential geometry, which experts expect will revolutionize the field. These techniques may prove sufficient to prove an even broader result in topology, known as Thurston's geometrization conjecture. For the subject of three-dimensional topology, this would be as fundamental a result as the discovery of the periodic table.

> Perhaps just as importantly, Perelman has introduced a raft of new techniques in differential geometry, which experts expect will revolutionize the field.

A Topological History Tour

What are these conjectures, and why are they so important? The answer calls for a little background.

In the nineteenth century, mathematicians came to grips with the mathematical nature of surfaces. They discovered a simple scheme for classifying surfaces using a single number, the so-called genus. Two surfaces are topologically identical—meaning it's possible to deform one into the other—if and only if they have the same genus.

The simplest example of a surface is, of course, the flat, Euclidean plane. But topologists prefer to work with surfaces that are closed and bounded—the term of art is "compact"—so the

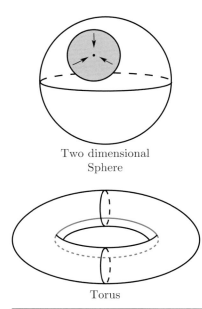

Two dimensional
Sphere

Torus

Figure 1. *A two-dimensional sphere is simply connected because any loop on the surface of the sphere can be tightened down to a point. On the other hand, a torus is not simply connected. The red loop shown here is caught around a neck of the torus and cannot be pulled any tighter.*

theory focuses instead on the sphere, which can be thought of as the plane with an extra "point at infinity" (see Figure "Stereographic Projection" in "Combinatorics Solve a Venn-erable Problem," p. 51). The sphere has genus zero. Roughly speaking, this means it has no "hole." Somewhat more precisely, it means that any closed loop drawn on the sphere can continuously shrink to a single point (see Figure 1 (top)).

The genus is always a non-negative integer, and there is a compact surface of each genus. A genus-1 surface has a single hole, and looks more or less like a doughnut. Such a surface is called torus. Because it has genus 1, not every closed loop drawn on the torus can shrink to a single point—a loop that goes around the hole, for example, cannot (see Figure 1 (bottom)). Surfaces of higher genus have correspondingly more holes, but the theory is still simple: If two compact surfaces have the same number of holes, then they are, topologically speaking, the same.

The mathematical nature of compact three-dimensional spaces, or 3-manifolds as they're called, is much more complicated. There is no simple number akin to the genus that distinguishes one 3-manifold from another. Researchers have instead developed a panoply of techniques for studying 3-manifolds.

The modern theory of 3-manifolds was initiated around 1900 by the French mathematician Henri Poincaré. Poincaré developed an algebraic theory in which closed curves in a 3-manifold are associated with elements in a mathematical group, called the manifold's *fundamental group*. It was a crucial insight for the emergence of topology as a viable branch of mathematics. Manifolds are floppy, hard to draw and hard to visualize. Algebra, on the other hand, is concise, precise, and lends itself to symbolic manipulation. If a manifold could be uniquely described by a group, the theory of manifolds would become vastly simpler.

Alas, such a neat correspondence was not to be. It's true that if two manifolds have different fundamental groups, the manifolds are topologically different. However, the converse is not true: Two manifolds can have the same fundamental group yet still be different.

In a paper published in 1904, Poincaré conjectured an important exception to this lack of a converse: If the fundamental group of a 3-manifold is trivial—which happens when every closed curve can shrink to a point—then that manifold is identical to the simplest possible 3-manifold, known as the 3-sphere.

The 3-sphere is a three-dimensional analogue of the ordinary two-dimensional sphere (sometimes called the 2-sphere). Just as the 2-sphere can be thought of as the Euclidean plane with an extra point at infinity, the 3-sphere can be thought of as Euclidean space with an extra point at infinity. It can also be viewed as the solution set in 4-dimensional space of the equation $x^2 + y^2 + z^2 + w^2 = 1$, just as the 2-sphere is the solution set in 3-space of the equation $x^2 + y^2 + z^2 = 1$ (and the "1-sphere," or circle, is the solution set in the plane of the equation $x^2 + y^2 = 1$—in general, the n-sphere is the solution set of the equation $x_1^2 + x_2^2 + \cdots + x_{n+1}^2 = 1$).

Actually Poincaré didn't phrase the conjecture as a conjecture. He posed it as a question: Is it possible for a 3-manifold to have trivial fundamental group without being identical with the 3-sphere? (The technical term for "identical" is "homeomorphic," which derives from the Greek for "same shape.") The difference between a conjecture and an open question is important. To pose a statement as a conjecture, one should have strong but not necessarily convincing evidence that it is true. Poincaré's caution may have been been prompted by his own bitter experience. Four years earlier he had stated, as a theorem, a stronger claim that turned out to be false; his 1904 paper retracted that claim and gave a counterexample. In his earlier work, Poincaré was studying a different and somewhat less powerful algebraic structure associated with manifolds, known as a homology group. He had falsely claimed that an n-manifold—for any dimension n—is homeomorphic to the n-sphere if it has the same homology groups as the n-sphere. His 1904 counterexample was a 3-manifold with the same homology groups as the 3-sphere but with a nontrivial fundamental group.

Poincaré was the first of many people to be fooled by the elusive 3-sphere. There were numerous attempts in the twentieth century to prove the Poincaré conjecture, and several claims to have done so. Like Fermat's Last Theorem (see "Fermat's Theorem—At Last!," *What's Happening in the Mathematical Sciences*, Volume 3), it came to occupy mathematicians' short list of notorious problems—seemingly simple problems that nevertheless mocked the attempts of even experienced mathematicians to solve them.

As mathematicians grew more interested in higher-dimensional manifolds (which cannot be visualized in our three-dimensional universe, but which are nevertheless every bit as "real" to a topologist), they naturally wondered whether the Poincaré conjecture could be extended to these manifolds as well. If an n-dimensional, compact manifold had a trivial fundamental group, was it necessarily an n-sphere?

At first blush, it might seem that the n-dimensional version of the Poincaré Conjecture must be much harder than the 3-dimensional version. After all, we can't even see what an n-dimensional space looks like. But the first major breakthrough on the Poincaré Conjecture came in 1960, when Stephen Smale of the Institute for Advanced Study and John Stallings at Oxford University independently proved that it was true for manifolds of 5 or more dimensions. Two decades later, in 1982, Michael Freedman, now at Microsoft Research in Bellevue, Washington, proved the conjecture for $n = 4$. As a result of the deep theorems of Smale, Stallings, and Freedman, mathematicians now knew how to identify the n-sphere in every case *except* the one Poincaré had originally asked about: the case $n = 3$.

Unfortunately, there seemed to be no chance that the proofs that worked in higher dimensions could somehow be adapted to three dimensions. The extra degrees of freedom available in higher dimensions sometimes permit techniques that don't work in lower dimensions. For example, in 4 and higher dimensions, every curve can be unknotted. But as we all know from experience, knots do exist in 3-space. Thus, for example, any

> Poincaré was the first of many people to be fooled by the elusive 3-sphere.

proof of the Poincaré Conjecture that involved untying knots would not be valid in 3 dimensions. An obstacle very much like this (only involving two-dimensional "knots") is a principal reason why Smale's and Stallings' proofs could not work in dimensions lower than 5.

Life in Three Dimensions

Three-dimensional manifolds have a very different "flavor" from high-dimensional ones. In the 1970s, William Thurston, then at Princeton University, realized that geometry plays a decisive role in low-dimensional topology, and in fact his geometric approach now dominates the subject. Thurston proposed what's called the geometrization conjecture. Very roughly speaking, geometry is topology with a notion of distance and angle; if topology is "rubber sheet" geometry, then geometry is "crystallized" topology. Thurston conjectured that every 3-manifold can be cut into pieces so that each piece will "freeze-dry" into a geometric structure associated with one of eight possible three-dimensional geometries. (See Figure 4, p. 11.) These geometries are well understood. In particular, only one of them has a trivial fundamental group: the 3-sphere.

Parts of Thurston's conjecture are incontrovertible. One is the existence of precisely eight geometries. A geometric structure, in Thurston's theory, is a compact three-dimensional space with a special way of measuring angles and distance known as a homogeneous Riemannian metric (named for the nineteenth-century mathematician Bernhard Riemann, who initiated the abstract study of manifolds). "Homogeneous" means the metric is the same at every point, much as homogenized milk has the same consistency throughout. The associated "geometry" refers to another three-dimensional space, usually non-compact, with the same metric, but with trivial fundamental group. (In the general theory of manifolds, every manifold, geometric or not, has a "universal covering manifold" which has trivial fundamental group. Because the universal covering manifold is not necessarily compact, it is not necessarily a sphere.) The geometric structure can be identified as a "quotient" of its associated geometry by a discrete group of motions, which is algebraically the same as the structure's fundamental group. Thus geometry, in Thurston's program, provides the link between topology and algebra.

The notion of geometrization is easier to understand in two dimensions, where something similar occurs. For surfaces, there are three different geometries: the sphere, the familiar Euclidean plane, and the less familiar, non-Euclidean, "hyperbolic" plane. The sphere is its own covering space; its group of motions consists of rotations. The Euclidean plane is the geometry for the torus: Since the torus can be viewed as a square with opposite edges identified, copies of it naturally tile the plane. The group of motions for the plane consists of translations and rotations.

Everything else—each surface of genus 2 and higher—is associated with the hyperbolic plane. The hyperbolic plane has a simple model: the interior of a circle. But instead of "lines" being lines in the usual, Euclidean sense, a hyperbolic line is a circular arc that meets the perimeter of the circle at right angles. This is what makes the hyperbolic plane non-Euclidean:

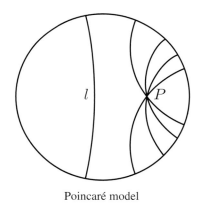

Poincaré model

Figure 2. *In the Poincaré disk model of hyperbolic geometry, the hyperbolic plane is represented by the inside of a disk, and lines are represented by circular arcs that are perpendicular to the boundary of the disk. Given any line l and any point P not on that line, there are many parallel lines to l through P, such as the four shown in this figure. This property distinguishes hyperbolic geometry from Euclidean geometry, where there is only one parallel to a given line through any point not on that line.*

Given a "line" and a point not on the "line," there is not just one "line" through the point that is "parallel" to the given "line"—there are, in fact, infinitely many "lines" through the point that don't intersect the given "line" (see Figure 2, opposite page).

The group of motions of the hyperbolic plane also consists of translations and rotations, but it permits tilings of the hyperbolic plane that are impossible in the Euclidean plane (see Figure 3, pages 8 and 9). This is what allows the hyperbolic plane to act as the geometry for higher-genus surfaces. For example, the hyperbolic plane can be tiled by right-angled octagons. (That is not possible in the Euclidean plane because right-angled octagons don't exist in Euclidean geometry.) Every motion of the hyperbolic plane that keeps the tiling intact corresponds to an element of the fundamental group of a two-holed torus. Thus the two-holed torus has a hyperbolic geometry.

A crucial aspect of these geometries is that they come with a way of measuring distance (called a Riemannian metric), and they endow their respective manifolds with a quality called curvature. The sphere has positive curvature, the Euclidean plane has zero curvature, and the hyperbolic plane has negative curvature. Because the metric is homogeneous, the curvature is the same at every point. For the sphere and hyperbolic space, it is usually normalized to 1 and -1, respectively.

The curvature of the sphere and the Euclidean plane are easy to understand. That of the hyperbolic plane is less clear. Curvature fundamentally has to do with what happens to nearby curves in a surface if they start out parallel and continue as straight as possible while remaining in that surface. In Euclidean geometry, they remain equidistant. In a surface of positive curvature, like the sphere, they move towards each other. (For example, think of two curves that start from the equator and head due south, moving as straight as possible *without leaving the surface of the sphere.* The two curves will always intersect at the south pole.) In hyperbolic geometry, on the other hand, curves that start out parallel tend to get farther and farther apart.

In sum, every surface is associated with one of three geometries. The 2-sphere (genus 0) is associated with itself. The torus (genus 1) has the flat geometry of the Euclidean plane. And surfaces of higher genus have a hyperbolic geometry. In effect, Thurston's geometrization conjecture says that much the same is true for 3-manifolds. Only there are eight geometries, not three. And the manifold may need to be cut into pieces first.

Three of the eight geometries are 3-D analogs of their 2-D counterparts. Spherical geometry is just the 3-sphere. Euclidean geometry is what you think it is. Hyperbolic geometry is a 3-D version of the hyperbolic plane, with the interior of the sphere acting as a model, much as the interior of a circle is a model for the hyperbolic plane. Two others are "product spaces" of the line with the sphere and the hyperbolic plane, respectively. These geometries are flat in one direction and curved in the other two. The final three geometries, called Solv, Nil, and SL$(2, R)$, are specialized; they occur rather rarely, but need to be taken in account for the sake of completeness. (This is perhaps a bit unfair. It's a little like dismissing 2 as an unimportant prime number just because even primes are so rare.)

> **In sum, every surface is associated with one of three geometries. The 2-sphere (genus 0) is associated with itself. The torus (genus 1) has the flat geometry of the Euclidean plane. And surfaces of higher genus have a hyperbolic geometry.**

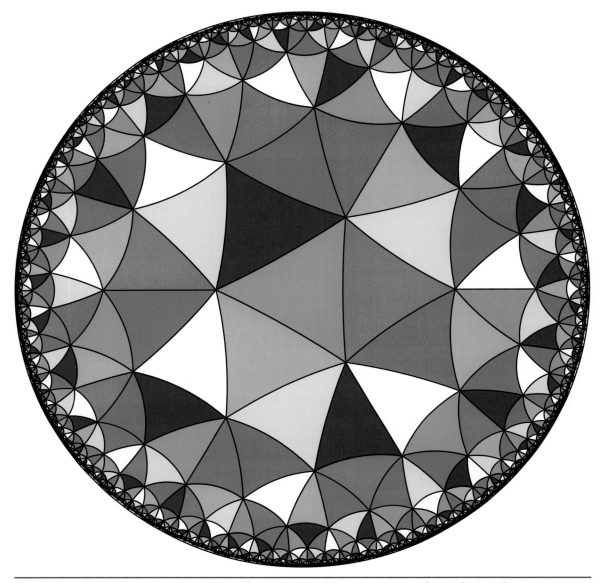

Figure 3a. *For caption, see Figure 3b on the next page. (Figure courtesy of Douglas Dunham.)*

Geometrization is a powerful concept because the structure imposed by a homogeneous Riemannian metric makes many topological properties much more accessible. In particular, the only geometry that permits fundamental groups of finite size is the spherical geometry— and the only one with trivial fundamental group is the 3-sphere itself. In other words, the Poincaré conjecture is an "easy" consequence of Thurston's geometrization conjecture.

Thurston and others have proved chunks of the geometrization conjecture. Thurston, for example, showed that a large class of 3-manifolds known as Haken manifolds (named after Wolfgang Haken, who is best known for his 1976 proof, with Ken Appel, of the famous four-color theorem) all have the geometry of hyperbolic space. (In a certain sense, "most" 3-manifolds are hyperbolic.)

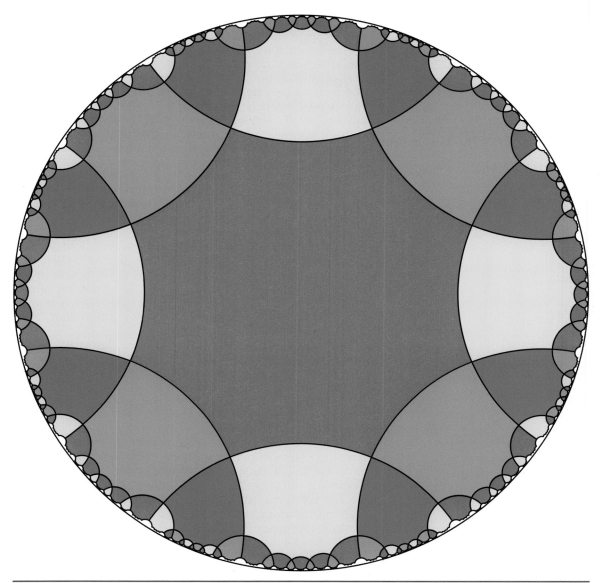

Figure 3b. *The hyperbolic plane has many more tessellations by regular polygons than the Euclidean does. These figures show two of them: a tessellation by equilateral triangles in which seven triangles meet at each vertex (a), and a tessellation by octagons in which four octagons meet at each vertex (b). These figures are not possible in Euclidean geometry because the sum of the angles meeting at each vertex of the tessellation would be greater than 2π radians. (Figure courtesy of Douglas Dunham.)*

Going with the (Ricci) Flow

In the early 1980s, Richard Hamilton, then at the University of California at San Diego, proposed a way of proving Thurston's geometrization conjecture using a concept that came to be called Ricci flow. Every manifold (in any dimension) comes equipped with some sort of Riemannian metric. The problem is, the metric may not be homogeneous. Hamilton's idea was to let the manifold "find its own way" to a homogeneous metric, by allowing the metric to evolve in a way very reminiscent of the way that heat flows. If heat flows unimpeded through a room

The "Ricci oven" does all the work!

with no heat sources or sinks, the temperature will eventually equilibrate. If the same thing could be done with a manifold, the curvature would eventually equilibrate—and the manifold would become homogeneous.

Heat flow is described by a partial differential equation that equates the rate of change of temperature in time to its variation (technically, its second derivative) in space. Hamilton was led to a similar equation for Riemannian metrics: Let the rate of change of the metric in time be proportional to a quantity called its Ricci curvature, named for the nineteenth-century Italian mathematician Gregorio Ricci-Curbastro. Ricci curvature is closely related to the notion of space-bending curvature in Einstein's general theory of relativity. Hamilton's Ricci flow equation has a very simple form: If $g = g(t)$ denotes the Riemannian metric at time t, then $\partial g / \partial t = -2R(g)$, where R is the Ricci curvature.

Hamilton's idea, in a nutshell, is that if a 3-manifold is in fact geometric, then no matter what Riemannian metric is placed on it to begin with, the Ricci flow equation will smooth it out so that the curvature is the same everywhere. In other words, starting from an amorphous, topological shape, the geometry of a 3-manifold will reveal itself through the limit of $g(t)$ as t goes to infinity.

John Morgan, a topologist at Columbia University, uses a cooking metaphor to describe the Ricci flow (and its connection with the heat equation): Imagine your manifold as a dollop of cookie dough, and the Ricci flow equation as an oven. The blob that goes into the oven has no particular shape, but what comes out is a nicely rounded—and crisp—cookie. The "Ricci oven" does all the work!

What if the manifold is a compound of two different geometries? Metaphorically speaking, what if it's an amalgam of cookie dough, pancake batter, and muffin mix? Ideally, the Ricci oven will cause the different pieces to separate, so that after a while the constituents are easily identifiable, with only a few tendrils keeping them connected—and eventually even the tendrils will snap or evaporate (see Figure 5, page 12).

Metaphor, however, is not proof. Hamilton's idea ran into theoretical obstacles. The main difficulties had to do with the nature of the tendrils—more technically known as singularities—that develop as pieces of a manifold try to separate. There seemed to be nothing preventing a thick tangle of singularities from developing, perhaps to the point that the metaphoric manifold would become all tendril and no cookie. Nor was it clear what kind of singularities were possible. In particular, Hamilton and others working on Ricci flow were not able to rule out a type of singularity they called the "cigar" (see Figure on page 2, "The Cigar"). The cigar, also known as Witten's black hole, is a rotationally symmetric solution of the Ricci flow equation for the (non-compact) Euclidean plane. In essence, it is a singularity that is perfectly happy "cooking" in the Ricci oven for all time. This would defeat the plan of having all the singularities break or evaporate.

Singularities don't necessarily show up at all. Hamilton showed the Ricci oven works perfectly on manifolds of positive curvature: Even if some parts are curved more than others,

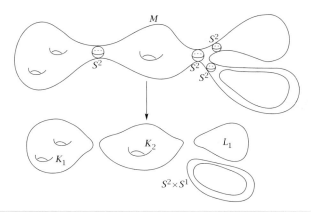

Figure 4. *Thurston's geometrization conjecture posited that any three-dimensional manifold can be decomposed into pieces that each have one of eight geometric structures. This figure illustrates pieces with four different structures: a two-holed torus would have a hyperbolic structure, a one-holed torus would have a Euclidean structure, a sphere would have a spherical structure, and a Seifert fibered manifold such as $S^2 \times S^1$ has a product structure. The conjecture allows some of the cuts to be made along tori (not shown here) rather than along spheres. For purposes of illustration the pieces are drawn as two-dimensional surfaces, but in reality they would be three-dimensional. (Reprinted from "Geometrization of 3-Manifolds via the Ricci Flow," by Michael Anderson, AMS Notices, February 2004, Figure 1, page 185.)*

the Ricci flow equation smooths things out so that the manifold asymptotically develops constant curvature. This showed that all such manifolds are geometric. Hamilton also proved that singularities take time to develop: If the metric is smoothly curved to begin, then it stays smoothly curved at least for a while. But it was clear that singularities do develop in the presence of negative curvature—and once they do, there was no telling what would happen.

Perelman changed all that. In a pair of "e-prints" posted online, the Russian mathematician laid out a program for dealing with the singularities that arise in Ricci flow. In the first paper, he analyzed the nature of the singularities and showed that the cigar (for example) does not occur. According to Morgan, experts realized very quickly that this part of Perelman's proof was for real, and this gave them motivation to tackle the much more difficult parts that followed. One of the key ingredients in the analysis is a notion of entropy for metrics. In thermodynamics entropy is a measure of disorder, which tends to increase over time. Perelman's metric entropy has a similar propensity to increase, which is what keeps the singularities under control.

Perelman's second paper shows how to deal with singularities as they arise. The basic idea is to cut out pieces of the manifold that are neighborhoods of developing singularities, glue in smooth pieces that are constructed by hand, and then let the Ricci flow continue to act on the modified manifold. One concern is that the cutting (topologists call it surgery, because it's actually not enough to cut—you also have to sew up the wound!)

> **Experts realized very quickly that this part of Perelman's proof was for real, and this gave them motivation to tackle the much more difficult parts that followed.**

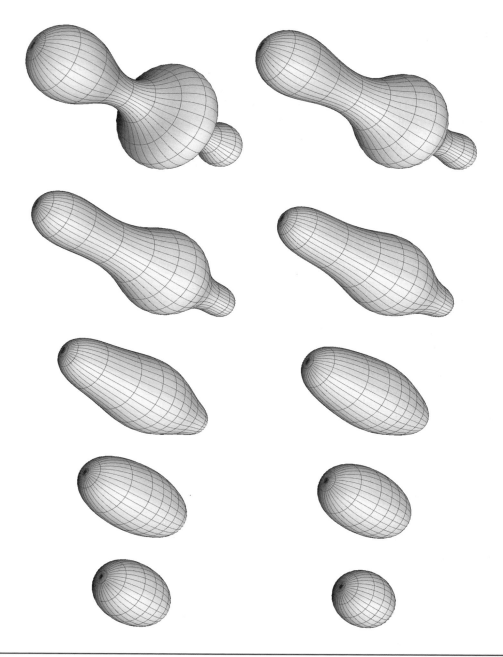

Figure 5. *Ricci flow on 2-dimensional manifolds tends to make the curvature more uniform, so they approach a sphere, a torus, or a multi-holed torus. This figure shows how an initially bumpy surface smoothes out to form a sphere. Ricci flow in 3 dimensions has the same tendency to even out the curvature, but it is greatly complicated by the fact that singularities such as narrow necks or horns can form and pinch off. (Courtesy of J. Hyam Rubinstein of the University of Melbourne and Robert Sinclair of the Okinawa Institute of Science and Technology.)*

could cause even more singularities to arise, so that one would wind up with an infinite number of pieces in a finite amount of time. Perelman overcame this with another argument reminiscent of the entropy idea. He showed that the surgery could be done in such a way as to reduce the volume of the manifold by a certain amount, $\epsilon(T)$. At the same time, the Ricci oven causes the volume to grow, just as a loaf of bread grows in a real oven. But if there were infinitely many surgeries in a finite time, they would subtract an infinite volume from the manifold—and the growth of volume due to the Ricci flow would not be enough to compensate. Therefore, only finitely many surgeries occur in any finite time interval. Thus the flow-with-surgery process can continue for all time, and that is enough time for the manifold to separate into pieces with distinct homogeneous geometries.

Perelman's work sets the stage for a proof of Thurston's geometrization conjecture. At the end of the second paper, he outlined how the argument goes and promised to give details in a third paper. That paper is yet to appear. However, Perelman posted a very short paper in July 2003 that gives a separate, simpler proof of the Poincaré conjecture, based on the results of the first two papers. At roughly the same time, Tobias Colding at the Courant Institute of Mathematical Sciences and William Minicozzi II at Johns Hopkins University gave another argument, also based on Perelman's two papers, to the same effect.

The simplicity of both proofs gave researchers confidence that the Poincaré conjecture at least was within their grasp—provided there were no lacunae in Perelman's main papers. That may seem like a simple matter of refereeing. But Perelman's work has kept experts busy for more than three years. The arguments are extremely technical, and there are a multitude of new ideas in the proof. Three separate teams of mathematicians have now produced book-length manuscripts explicating Perelman's proof. One team, Bruce Kleiner and John Lott of the University of Michigan, posted their work on the Internet as they progressed. The second team, Huai-Dong Cao of Lehigh University and Xi-Ping Zhu of Zhongshan University in China, published their paper in the June 2006 issue of *Asian Journal of Mathematics*. The third team, Morgan and Gang Tian of the Massachusetts Institute of Technology, plans to publish their manuscript as a book. Morgan says that he is convinced, but that does not mean that the story is over. "The experts are very optimistic but cautious. If this were an ordinary problem, we would have been satisfied two years ago. We have only continued [to question it] because it's the Poincaré Conjecture and it has such a long history of mistakes, even by very good mathematicians."

To be eligible for the Millennium Prize, a solution of the Poincaré Conjecture must be published in a refereed journal "or other such form as the Science Advisory Board [of the Clay Mathematics Institute] shall determine qualifies" and to survive two years of further scrutiny after that. Perelman himself has been conspicuously silent since releasing his preprints, and has never submitted them to a journal. However, James Carlson, the president of the Clay Mathematics Institute, has confirmed that the three exegeses of Perelman's work satisfy the publication requirement, and the two-year clock is now "ticking." Thus it seems likely that mathematicians' first great breakthrough of the third millennium will be ready to enter the history books sometime in 2008 or 2009.

> **"If this were an ordinary problem, we would have been satisfied two years ago."**

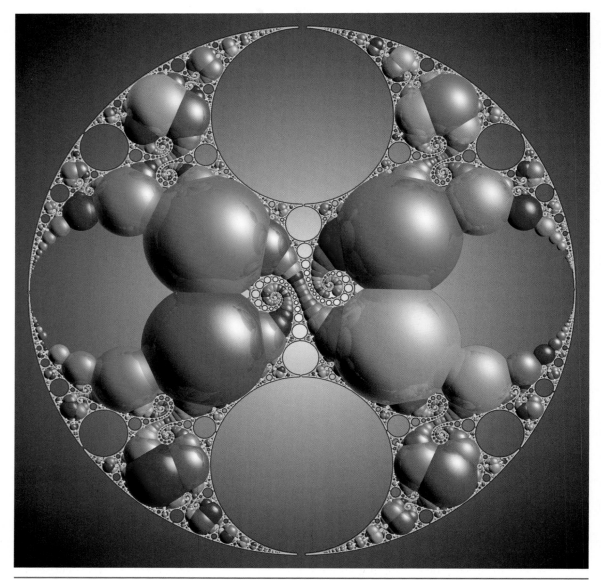

Hall of Mirrors. *Spectacular images result from plotting the action of a Kleinian group in 3-dimensional space. To a topologist, this image represents the outside, or convex hull, of a 3-dimensional "hall of mirrors." Each individual room in the hall of mirrors corresponds to a 3-dimensional manifold. In a case like this one, where the walls of the hall of mirrors form visible spheres, the manifold is said to be "geometrically finite," and its ends are tame. Until recently topologists did not know if "geometrically infinite" manifolds, where the spheres shrink down to an infinitely fine froth, also had tame ends. (© JosLeys.)*

Classifying Hyperbolic Manifolds— All's Well That Ends Well

Dana Mackenzie

WHILE THE STATUS of one famous unsolved problem in topology remained uncertain, topologists conquered four other major problems in 2004. The four theorems that went into this unique sundae—the Ending Laminations Conjecture, the Tame Ends Conjecture, the Ahlfors Measure Conjecture, and the Density Conjecture—were not household names, even among mathematicians. Yet to topologists who study three-dimensional spaces, they were as familiar (and went together as well) as chocolate, vanilla, hot fudge sauce and a maraschino cherry.

The Poincaré Conjecture (see "First of Seven Millennium Problems Nears Completion," p. 2) can be compared to the quest for the Loch Ness Monster: Whether you discover it or prove it doesn't exist, fame and glory are sure to follow. On the other hand, if topologists spent all their time hunting Loch Ness monsters, the subject would never advance. Someone has to study and classify the spaces that are already known. That is what the Ending Laminations Conjecture and Tame Ends Conjecture are all about. They complete the taxonomy of a very broad category of three-dimensional spaces, called hyperbolic 3-manifolds. Though they may not be quite as glamorous as deep-sea monsters, they are much more numerous.

"I've been working on the Poincaré Conjecture for most of my career," confesses Danny Calegari of Caltech. But when rumors started circulating in 2003 that Grisha Perelman had solved it, Calegari decided he had better look for a new problem to work on. Dave Gabai of Princeton University suggested that they could team up to work on the Tame Ends Conjecture, which Gabai had already been thinking about since 1996. It was a problem that had been bouncing around for thirty years, since it was first proposed by Al Marden of the University of Minnesota. At the time he proposed it, Marden says, "It was pie in the sky. No one had the vaguest idea how to prove it." Gabai had already tried to prove the Tame Ends Conjecture once, with Michael Freedman. Although their program had not worked—they found a counterexample to their original idea—Gabai wanted wanted help with a promising variant of Freedman's approach, called "shrinkwrapping."

Meanwhile, Ian Agol of the University of Illinois at Chicago was going through a very similar process. He, too, was frustrated with trying to understand Perelman's work on the Poincaré Conjecture, but had an idea how to prove the Tame

Danny Calegari. *(Photo courtesy of Danny Calegari.)*

Dave Gabai.

Ian Agol. *(Photo courtesy of Ian Agol.)*

Jeffrey F. Brock. *(Photo courtesy of the University of Texas at Austin.)*

Ends Conjecture. Both groups—Gabai and Calegari together, and Agol working alone—finished their work at roughly the same time.

The proof of the Ending Laminations Conjecture, on the other hand, resulted from an intensive multi-year effort. The classification of hyperbolic manifolds depends on identifying a "feature" that topologists didn't even know about until William Thurston, then at Princeton University, discovered it in the 1970s. Thurston sketched out a rough idea for how this part, called an ending lamination, could be used to identify hyperbolic manifolds, but left huge gaps for other mathematicians to fill. The first step was a rigorous proof that ending laminations even existed—a proof that was provided by Francis Bonahon of the University of Southern California. The second step, using the laminations as a scaffolding to piece together a whole manifold, looked so difficult that only one mathematician seriously believed he could do it. Yair Minsky of Yale University began working on the Ending Laminations Conjecture shortly after receiving his Ph.D. from Princeton in 1989, and turned it into his career project. Working with a steady stream of collaborators—most notably Howard Masur, Dick Canary, and Jeff Brock—he gradually stitched together pieces of the proof until, by December 2004, he was finally ready to pronounce it complete. "In my mind," says Bonahon, "the real achievement is the work of Brock, Canary and Minsky. Agol, Calegari, and Gabai put a cherry on top—although a very impressive cherry." Minsky is somewhat more modest: "I'm glad that Francis feels this way, but personally I am happy with the role of one out of several cherries on a very nice sundae." Marden, on the other hand, dismisses all talk of sundaes and emphasizes the great importance of both theorems. "The most fundamental [result] is the tameness," he says. "For without knowing that, both ending laminations and density would be known only for 'tame' manifolds. There would be no way of knowing how encompassing a class this is." On the other hand, he adds, "The deepest of the theorems, the one that was hardest and required the deepest penetration into the hyperbolic geometry is ending laminations."

The final two ingredients in the sundae, the Density Conjecture and the Ahlfors Measure Conjecture, stem from an earlier era in the study of hyperbolic geometry, before Thurston revamped the subject and made it part of topology. In the 1960s, such mathematicians as Lars Ahlfors and Lipman Bers had taken a completely different approach to the subject, viewing it as a branch of complex analysis (the study of functions of complex numbers). They were interested in certain symmetries of complex functions known as Mobius transformations. Their original project was to classify groups of Mobius transformations, known as *Kleinian groups* (to be described below). But as often happens in mathematics, the problem could not be solved in the framework it was originally conceived in; it was only after Thurston rephrased the problem in terms of hyperbolic manifolds that major progress became possible. "We didn't even know how little we knew until Thurston," says Marden.

Ahlfors found that there were certain nicely behaved Kleinian groups that could be classified with the tools of complex analysis, but there were others that could not. The relation between them was very much like the relation between rational and irrational numbers. The decimal expansion of rational numbers marches in a precise cadence forever; irrational numbers may act rational for a long time, but they always break stride and follow their own drummer eventually. The Density Conjecture was a sort of formalization of this metaphor. It states that any hyperbolic manifold (or any group of Mobius transformations) can be approximated "arbitrarily well" by a well-behaved one.

What are hyperbolic manifolds, what are Mobius transformations, and what do they have to do with one another? A topologist would answer this question in one sentence: a hyperbolic manifold is a quotient of hyperbolic space by a discrete group of Mobius transformations. (He would mumble a few more words that sound like "orbifold" and "finitely generated," but non-topologists are free to ignore the mumbled part.) As usual in mathematics, quite a bit of explanation is needed to understand the one-sentence answer.

A three-dimensional manifold is the mathematician's way of describing the universe we live in. On a small scale, a manifold looks exactly like our familiar three-dimensional space, possibly with subtle distortions due to curvature. In places where the curvature is positive, parallel lines (which you can think of as light rays) bend toward each other as if they were passing through a lens; where the curvature is negative, light rays bend away from each other. The simplest kind of universe is one where the curvature is the same at every point. Such a manifold is called *hyperbolic* if the curvature is negative. Thurston was the first mathematician to realize that the vast majority of interesting three-dimensional manifolds are hyperbolic, or can be put together out of hyperbolic pieces.

On a large scale, manifolds can connect up in complicated ways, with wormholes and handles that may or may not exist in our universe. They may also have "ends," which are tubelike or flaring conelike structures that extend infinitely far into the distance. The Tame Ends Conjecture, as its name implies, states that the ends of hyperbolic manifolds cannot get very complicated. The Ending Laminations Conjecture complements this by stating that the geometry of tame ends completely controls the rest of the space. If you were the god of a hyperbolic universe with tame ends, you would find it very compliant. You would be able to change its shape just by tugging on its ends, and people living inside the universe would not have any say at all. (See Figure 1.)

An extreme case is a hyperbolic universe of finite size. Such a universe has no ends to tug on, and thus the universe can have only one possible shape. To be more precise, each possible *topology* has one unique *geometry* of constant negative curvature. That is exactly the content of the Mostow Rigidity Theorem, proved in 1973 by George Mostow, which Masur calls "the most influential piece of mathematics in geometry in the last 35 years." One of the main motivations behind the Tame Ends Conjecture and the Ending Laminations Conjecture was to explain how rigidity translates to hyperbolic universes that are infinitely large.

Yair Minsky. *(Photo courtesy of Yair Minsky.)*

Dick Canary. *(Photo courtesy of The University of Michigan Photo Services.)*

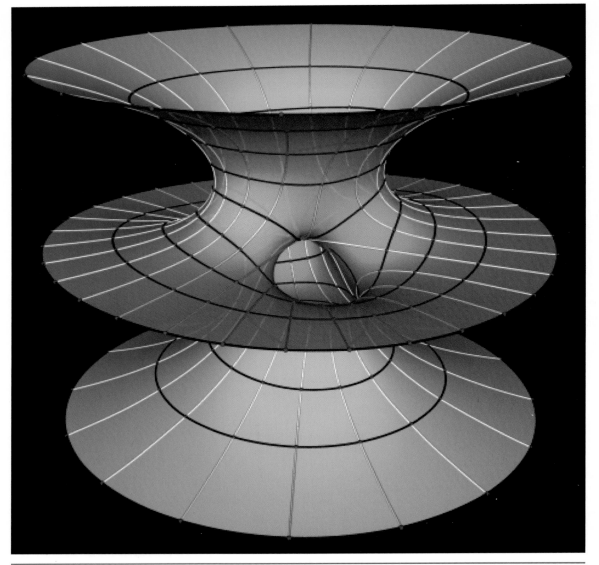

Figure 1. *An "end" of a non-compact manifold can be thought of as a piece of the manifold that stretches out to infinity. This image shows a famous surface, called the Costa-Hoffmann-Meeks surface, that has three ends—the three roughly planar pieces that stretch out to infinity. (Figure used with permission of Matthias Weber.)*

As implied by the one-line definition of hyperbolic manifolds given above, there is a second way of looking at them that doesn't involve universes, light rays, or benevolent gods. It involves Mobius transformations instead.

Most math majors first encounter Mobius transformations in a course on complex functions. In complex analysis, the Euclidean plane is represented by complex numbers $z = x + iy$, rather than by ordered pairs (x, y). Functions that map the plane to itself without distorting angles turn out to be particularly important and easy to represent with complex numbers; such functions are called *conformal*. The most general kind of conformal map that doesn't "miss" any points on the plane is called a Mobius transformation, and it has a particularly simple formula: $f(z) = (az + b)/(cz + d)$. (Here the variable

z and the constants a, b, c, d are all complex numbers.) Again, the purists would mumble words like "orientation-preserving" and "point at infinity," but you are free to ignore them unless you are being tested.

The wonder of Mobius transformations is that there is also a geometric or pictorial way of looking at them. You can create any Mobius transformation by doing a series of reflections in the complex plane, using either an ordinary flat mirror or a curved, circular one.

Once you have mirrors, you can put them together to create a "hall of mirrors" effect. In an ordinary room whose walls are flat mirrors, you can see copies of yourself in every direction you look. They form a crowd of clones (half of them mirror-images, the others exact copies of you) that seem to recede into the infinite distance. But what happens if you replace the flat mirrors with circular ones? Felix Klein, a nineteenth-century German geometer, discovered an amazing fact. You would still see infinitely many clones of yourself. But they would *not* appear to recede into the infinite distance. If there were three circular mirrors, the reflections would accumulate along a circle; and if there were four or more mirrors, the reflections would accumulate along a *limit set*, an extraordinarily jagged curve that Klein tried and failed to sketch. Nowadays we know why he found it so hard: It was the first appearance in the mathematical literature of a *fractal*.

In the computer age, we are no longer encumbered by the limitations of human imagination and draftsmanship. Pictures of fractals have become iconic images of the computer age, and pictures of Kleinian limit sets are among the most beautiful. A selection of images corresponding to different mirror sizes and placements is shown here. (Note that the circular mirrors have been replaced in some examples by spheres.) Every Kleinian limit set has its own unique style and beauty. Technically, the hall of mirrors is called a *Fuchsian group* (if the mirrors are circles in the plane) or a *Kleinian group* (if they are spheres in space). (See Figure 2.)

You can play around with Fuchsian groups and Kleinian groups to your heart's content without ever mentioning a single word of topology. However, the more you play, the more you will get the feeling that there is some organizing principle behind them. To understand that principle, you need topology and the concept of hyperbolic manifolds. (See Figure "Hall of Mirrors" on p. 14.)

As the one-line definition of hyperbolic manifolds said, you get a hyperbolic manifold by taking a quotient of the hyperbolic plane (or space) by a Fuchsian (or Kleinian) group. What does this mean?

Henri Poincaré—yes, the same man who came up with the Poincaré Conjecture—realized that if you are standing in a hall of mirrors, you have no way of telling whether you are in an infinite universe with infinite copies of yourself, or just a single room with mirrors on the walls. You might argue that there is one surefire if somewhat painful way to tell the difference: If you are in a single room, if you keep walking in one direction you will eventually bump into a wall. However, mathematically it is easy enough to arrange for the mirrors to be permeable, so that you can step right through them like Alice through the

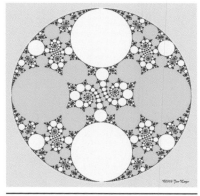

Figure 2. *This beautiful fractal portrays the action of a Kleinian group (a discrete group of Mobius transformations) on the complex plane. Any white disc can be mapped to any other white disc by one of the transformations in the Kleinian group. The group also maps the shaded region to itself. The fractal boundary between the two regions is the limit set of the group. The shaded region can also be thought of as the union of an infinite collection of "rooms" in a "hall of mirrors." An observer inside the shaded region cannot tell whether she is in a single room with mirrors or a gallery with infinitely many rooms. Even if she is in a single room, she will see infinitely many reflections of herself clustered in the distance, and the apparent shape traced out by these reflections will be identical to the Kleinian limit set. (© JosLeys.)*

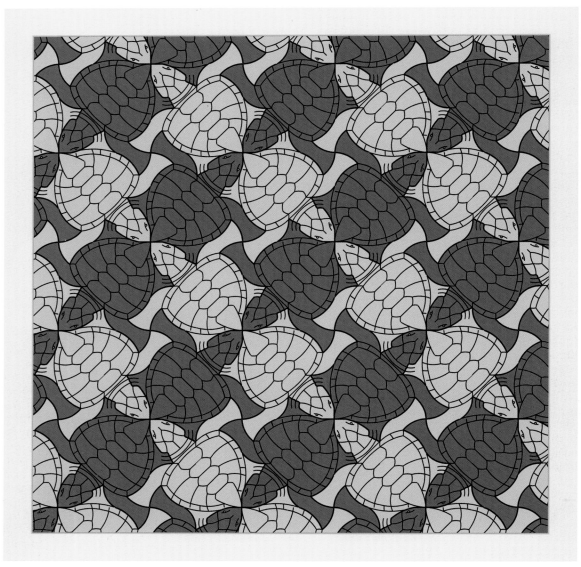

Figure 3a. "Sea Turtles 1". *For caption, see Figure 3b on next page. (© JosLeys.)*

looking glass. If the universe is infinite, you would keep going into a new room that looks like the one you just left. If the universe is finite, you would re-emerge somewhere else in your original room. Either way, you cannot tell the difference. You are living in . . . the Quotient Zone. (Cue *Twilight Zone* music.)

A beautiful two-dimensional example of this phenomenon can be seen in the graphics of Jos Leys, a Belgian artist who is strongly inspired by the late M.C. Escher (see Figure 3). In his picture "Sea Turtles," we apparently see an entire plane with infinitely many turtles. Or do we? We could get exactly the same picture by placing mirrors along the backbones of four of the turtles, forming a square. In the Quotient Zone, there are only four "real" half-turtles, and everything else is a reflection of a reflection of a reflection. An example in hyperbolic geometry is provided by Leys' picture, "Fish." Now the backbones of six adjacent fish form a right-angled hexagon (a figure that exists

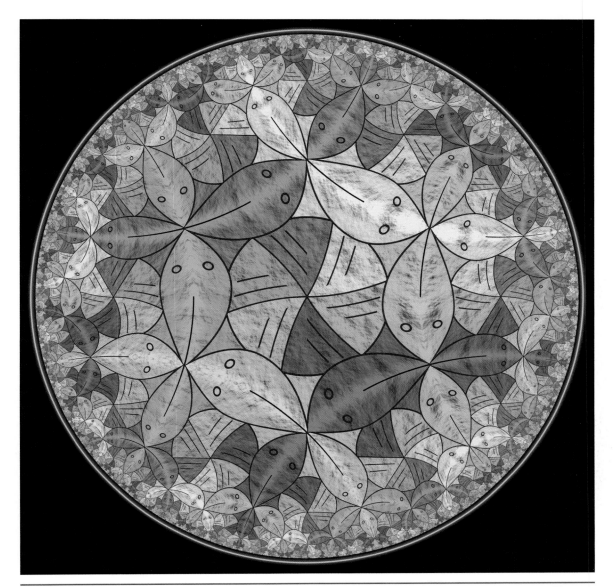

Figure 3b. "Fish". *Two tessellations by Jos Leys illustrate the difference between Euclidean and hyperbolic geometry. "Sea Turtles 1" is a tessellation of the Euclidean plane. Every turtle has the same size and shape. "Fish" is a tessellation of the hyperbolic plane. Every fish has the same size and shape, when measured according to the rules of hyperbolic geometry. In each image, the backbones of the animals reveal the underlying geometry of the space they live in. The backbones of the turtles form a square grid, while the backbones of the fish form a pattern of right-angled hexagons. Right-angled hexagons exist only in hyperbolic geometry, just as squares exist only in Euclidean geometry. (©JosLeys.)*

only in hyperbolic geometry). Are there infinitely many fish, or only six half-fish? I'm not telling. Sometimes it is more convenient to look at it one way, and sometimes it is more convenient to look at it the other.

The first way is better if you want to visualize the ends of a hyperbolic manifold. In this view, identical copies of the manifold fill an entire disk (in two dimensions) or ball (in three dimensions), just as Leys' pictures are filled with turtles and fish. The disk or ball is called the *Poincare disk model* of the *hyper-*

bolic plane (or *hyperbolic space*). Ends arise if the hyperbolic manifold (or, equivalently, any one room of the hall of mirrors) extends all the way out to the boundary of the Poincaré disk. This makes the manifold infinitely large because of the way that distances are measured in hyperbolic geometry. The center of the Poincaré disk is infinitely far from its boundary, as you can see from looking at Leys' picture. Each of the fishes in the picture is defined to be the same size in hyperbolic geometry, and you have to pass by infinitely many fishes to reach the boundary. In other words, you have to travel an infinite distance.

So far there has not been much of a difference between 2- and 3-dimensional hyperbolic manifolds, aside from the sub-

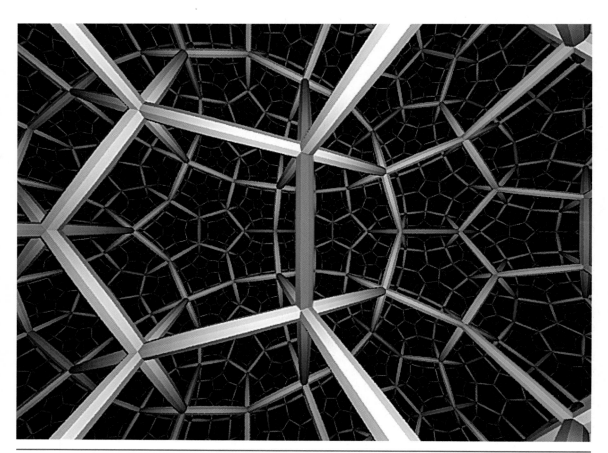

Figure 4. *The view from "inside" a hyperbolic 3-manifold. The manifold has finite volume and size, but it appears infinite for the same reason that a room with mirrors does. The adjacent "rooms" in this picture are really the same room, apparently rotated and translated. In this image the manifold is compact (i.e., the rooms have finite size), but the same methods can be used to study non-compact 3-manifolds as well. (Figure courtesy of The Geometry Center, University of Minnesota, © 1990. All rights reserved.)*

stitution of "space" for "plane" and "ball" for "disk." As in the 2-dimensional case, 3-dimensional manifolds can be pictured as individual rooms in a hall of mirrors, with the various reflections of the room apparently filling out all of hyperbolic space (see Figure 4). However, when we start talking about ends, a dramatic difference appears. In 2 dimensions, the ends are just arcs on the boundary circle, and the limit set of the Fuchsian group is the collection of endpoints of those arcs. But in 3 dimensions, we have a boundary *sphere*. The hyperbolic manifold corresponds to a "room" inside the sphere, with exotic matching rules on the walls. (Remember that when you are in the Quotient Zone, walking through one wall causes you to re-emerge from a different wall in the same room; the matching rules say which one.) The ends are polygons on the boundary sphere, which have matching rules of their own. These rules glue the edges together in such a way that the polygons become 2-dimensional surfaces (such as doughnuts or pretzels). The Kleinian limit set is a gorgeous fractal, where infinitely many of the end-polygons cluster together. The limit set and the ends are complementary; everything on the boundary sphere that is not part of an end of the hyperbolic manifold is part of the Kleinian limit set. (See Hall of Mirrors, p. 14.)

Back in the 1960s, Ahlfors showed that if the number of walls of any "room" is finite, then the shape of the end polygons translates directly to a unique hyperbolic geometry on the "rooms" and the hyperbolic manifold they represent. In fact, according to Marden, Ahlfors initially thought that his theorem settled the question once and for all. Topologists now call it the "geometrically finite" case. But it turned out that in three dimensions (unlike two) the "rooms" of the hall of mirrors could contain infinitely many sides. In this case the Kleinian limit set suddenly changes its nature from an elegant collection of filigrees into a violently jagged, tentacled monster that gobbles up almost the entire boundary sphere. (This sudden transformation is the essence of the Ahlfors Measure Conjecture, the fourth theorem in the "sundae.") Even with a computer, it is hard to draw an accurate picture of it. It was this monster—the case of the "geometrically infinite" ends—that required the efforts of six people (Agol, Calegari, Gabai, Brock, Canary and Minsky) to tame and assign its proper taxonomy.

At this point the saga of the hyperbolic manifolds splits, like the *Lord of the Rings* saga, into two parties. Agol, Calegari and Gabai chose to work on the taming of the ends.

In the geometrically infinite case, the ends of the hyperbolic manifold actually do not make it to the boundary sphere. The reason is that the Kleinian limit set takes away so much of the sphere that there is nothing left for the ends. Unfortunately, this completely negates the usefulness of the first way of looking at Kleinian groups: we can't very well describe how the ends control the geometry of the hyperbolic manifold if we can't even find the ends. This is, perhaps, why Ahlfors was stymied by the geometrically infinite case.

However, the ends are still there, and you can still study them by going to the Quotient Zone point of view. In this interpretation, the hyperbolic manifold (the "room" in the hall of mirrors) has a long, undulating tunnel that reaches out toward infinity.

At this point the saga of the hyperbolic manifolds splits, like the *Lord of the Rings* saga, into two parties.

The end of the manifold is *tame* if it is topologically the same as a straight tunnel with no undulations. The reason the ends are hard to tame is that the tunnel may twist infinitely often, and in the course of this twisting become so knotted that it can no longer be straightened out. Somehow the constant negative curvature of the hyperbolic manifold (remember that?) must prevent this kind of infinite twist. The proof by Calegari and Gabai involved a technique they invented, called "shrinkwrapping." They showed that you could take a short loop that circles the tunnel and "shrinkwrap" it—that is, you push a surface outward from the "core" of the manifold until it gets caught on the loop, then pull it tight. A simple but elegant geometric argument (which works only because of the manifold's constant negative curvature) shows that the area of the shrinkwrap does not increase as the loop moves farther and farther out along the tunnel. But that means that the end cannot be infinitely twisted and knotted because if it were, the shrinkwrap for the more distant loops would get tangled up with the closer loops, and the areas would have gotten bigger and bigger. Therefore, the end must have been tame to begin with!

Meanwhile, Minsky and his troops were working to understand the structure of ends once they had been tamed. Once you know the end has the structure of a straight tunnel, you can take any cross-section of that (three-dimensional) tunnel and get a two-dimensional surface—for example, a torus with a puncture in it. Every cross- section will be the same, so it is possible to pick one cross-section as a model for all the others.

Next, Minsky (following Thurston's idea) considered a sequence of loops heading out the end of the tube—the same loops that Calegari and Gabai used in their shrinkwrapping argument. He slid these loops back to the reference surface. But when you slide a curve in hyperbolic geometry, it invariably becomes longer. (This is because of the defining property of negative curvature: parallel lines diverge. Hence the ends of a line segment get farther and farther apart when you slide them in parallel directions.) Thus if your first curve wraps around the reference surface once, the second one will be longer and it will have to wrap around it more than once—or perhaps once longitudinally and once latitudinally, creating a "barber pole" effect.

As the curves get longer and longer, one of two things could happen. They could start to retrace the same path (an easy way to increase distance—just travel the same path over and over), or they could fail to close up ever, and just create a denser and denser sequence of barber pole stripes. In the limit, the barber pole stripes become infinitely long and infinitely dense—and that limit is called a *lamination* of the reference surface. It was this structure, Thurston believed, that contained all the relevant information about the shape of the end, and completely determined the geometry of the rest of the manifold. This assertion became known as the Ending Laminations Conjecture.

One challenge for Minsky was to figure out how different ends communicate with each other—because if there is more than one end, each of them will play a role in shaping the manifold. The simplest, and first, case that he worked on was that of a manifold with two ends, each with the same reference surface. For example, if the reference surface were a punctured torus, the manifold would be a thick rubber inner tube with a nail hole running from the outside surface to the inside surface. One might call it a punctured tire. The outside of the inner tube would have one ending lamination and the inside, which you cannot see, would have another.

In this example, a barber pole stripe that winds around the reference surface (the punctured torus) p times longitudinally and q times latitudinally corresponds to a rational number, p/q. Ending laminations correspond to irrational numbers, such as the golden ratio (1.618033985...). To move from one ending lamination to another, one proceeds by a series of hops from one "nearby" rational number to the next. In the case of the golden ratio, one might hop along Fibonacci ratios—from 13/8 (1.625) to 8/5 (1.6) to 5/3 (1.666...) to 3/2 (1.5) and so on, gradually moving away from the golden ratio and towards the irrational number representing the second ending lamination. (See Figure 5, next page.)

At each step, hopping from one ratio to the next corresponds to moving from one barbershop spiral to another one that does not intersect it. (They only intersect at the puncture point, which has been removed for precisely that reason.) It is hard for two different spirals not to intersect, and so the information on which spirals do not intersect turns out to be crucial in reconstructing the geometry of the manifold. Minsky showed (in about 1994) that if you go all the way from one lamination to another (which requires an infinite number of "hops") you will get all the information you need to build a complete model of your manifold. At first it is not a perfect model, but sort of a stitched-together Frankenstein-like version of it. But in their 120-page paper, released as a preprint in 2004, he showed with Brock and Canary how to smooth out the stitching so that you end up with an exact copy of your original manifold, with the desired ending laminations on both ends. Most impressively, they described how to do it for all manifolds, not just the punctured-tire manifold described here. Thus, he concluded, Thurston was right. The ending laminations for all geometrically infinite ends, plus the shape of any geometrically finite ends, completely describe the geometry of any hyperbolic manifold. (One more time there are some mumbled words that you can ignore, concerning a "finitely generated fundamental group.")

What can you do for an encore after bringing a 15-year research program to a close? "In this field we've been obsessed with manifolds of infinite volume for a long time," says Brock. "One thing that we're doing now is taking the technology of these proofs and translating it back to manifolds of finite volume."

For example, Brock says that any three-dimensional hyperbolic manifold can be sliced into two pieces in a special way called a *Heegaard splitting*. The two pieces are topologically identical, and they look like balls with lots of handles of different shapes attached. The trouble with the Heegaard splitting is that the pieces have open, raw edges—the places where the original manifold was cut open. Topologists did not have the tools to relate combinatorial information about the edges to the geometry of the pieces. But now they do. Manifolds with open edges are a lot like manifolds with ends (although the edges

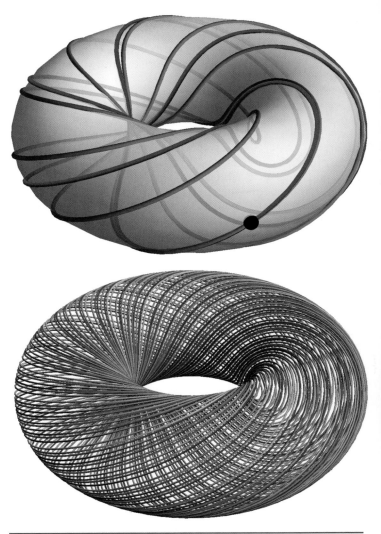

Figure 5. *A lamination on a torus is obtained as a limit of torus knots, i.e., curves that wind around the torus and close up. Here, the blue curve circumnavigates the torus 3 times in latitude while going around 5 times in longitude. Thus it has "slope" 5/3 = 1.666. . . . Similarly, the red curve has "slope" 8/5 = 1.6. These two curves intersect only at the puncture point (black dot). The multicolored curve has an irrational "slope" of 1.618. . . . It winds around forever without ever closing up or intersecting itself. This is called a lamination. (Graphics created by Michael Trott using Mathematica.)*

do not extend out to infinity). The region around the edges can be equipped with a curve complex, the very same type of scaffolding that Minsky used to generate a blueprint of a manifold with ends. "This is neat because there are people who understand the topology of Heegaard splittings in a very deep way," says Brock. "Our challenge is to relate this topology to the curve complex."

Minsky also is optimistic that there is a lot still to do. The fact that you have a description of a manifold's shape does not mean you have a *convenient* description. An analogous situation can be found even in high-school Euclidean geometry. The three side lengths of a triangle "completely describe" its shape—but that doesn't necessarily make it easy to answer concrete questions about the triangle. For example, geometers still do not know which triangles have periodic billiard-ball trajectories—even though that information must somehow be encoded in the side lengths. Minsky says that topologists are still a long way from understanding how the ending laminations, which "completely describe" a hyperbolic manifold, relate to other geometric properties of the manifold.

As for Marden, he is thrilled to see his conjecture proved after thirty years. "So much has happened," he says. "If you ever questioned whether there is progress being made in mathematics, this is a very clear case. These proofs could not have been done thirty years ago."

> **"If you ever questioned whether there is progress being made in mathematics, this is a very clear case. These proofs could not have been done thirty years ago."**

Tombstone. *The tombstone of Ludolph van Ceulen in Leiden, the Netherlands, is engraved with his amazing 35-digit approximation to pi. Notice that, in keeping with the tradition started by Archimedes, the upper and lower limits are given as fractions rather than decimals. (Photo courtesy of Karen Aardal.* ©*Karen Aardal. All rights reserved.)*

Digits of Pi

Barry Cipra

THE NUMBER π—the ratio of any circle's circumference to its diameter, or, if you like, the ratio of its area to the square of its radius—has fascinated mathematicians for millennia. Its decimal expansion, 3.14159265..., has been studied for hundreds of years. More recently its binary expansion, 11.001001..., has come under scrutiny. Over the last decade, number theorists have discovered some surprising new algorithms for computing digits of *pi*, together with new theorems regarding the apparent randomness of those digits.

In 2002, Yasumasa Kanada and a team of computer scientists at the University of Tokyo computed a record 1.2411 *trillion* decimal digits of π, breaking their own previous record of 206 billion digits, set in 1999. The calculation was a shakedown cruise of sorts for a new supercomputer. That's one of the main reasons for undertaking such massive computations: Calculating π to high accuracy requires a range of numerical methods, such as the fast Fourier transform, that test a computer's ability to store, retrieve, and manipulate large amounts of data.

The first rigorous calculation of π was carried out by the Greek mathematician Archimedes in around 250 BC. Archimedes did not have the decimal system at his disposal. Instead he worked with fractions. In his treatise *The Measurement of the Circle*, only parts of which have survived, Archimedes showed that π is greater than $3\frac{10}{71}$ but less than $3\frac{1}{7}$. (The latter value, also written as 22/7, is one many people remember from school as "the" value of π.) In modern, decimal terms, these values are correct to two decimal digits: $3\frac{10}{71} = 3.14084\ldots$ and $3\frac{1}{7} = 3.14285\ldots$.

Archimedes's method is based on polygonal approximations to the circle. The basic idea is to repeatedly double the number of sides in the polygon. Archimedes in effect worked out the relationship between the perimeter of the new polygon (with twice as many sides) and the perimeter of the old one (see Figure 1, next page). As the number of sides increases, the polygon becomes more and more indistinguishable from a circle, and the perimeter of the polygon becomes closer and closer to the circumference of the circle. If the polygon is inscribed in the circle, its perimeter will be slightly too short; if it is circumscribed around the circle, the perimeter will be slightly too long. Archimedes started with a hexagon and doubled the number of sides four times, so his under- and overestimates for π were based on a polygon with 96 sides.

Over the centuries, other mathematicians extended Archimedes's calculation with more and more doublings. This approach peaked around 1600, when the Dutch mathematician Ludolph van Ceulen used 60 doublings of an inscribed square

In 2002, Yasumasa Kanada and a team of computer scientists at the University of Tokyo computed a record 1.2411 *trillion* decimal digits of π, breaking their own previous record of 206 billion digits, set in 1999.

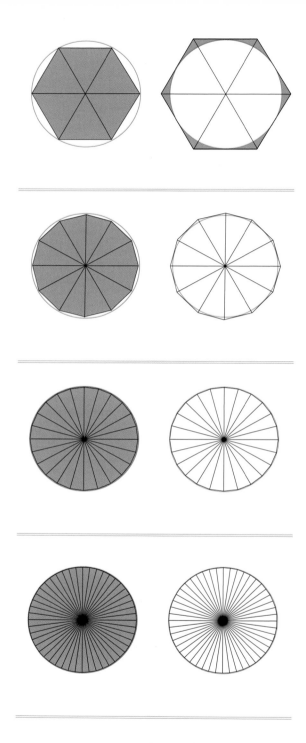

Figure 1. *Archimedes computed one of the first good approximations to pi by starting with a hexagon inscribed in a circle or circumscribed about it. The perimeter of the hexagon is easy to compute, and gives a rather inaccurate approximation to pi. Each time the number of sides is doubled, the approximation gets better. Archimedes used basic trigonometry to figure out how to compute the perimeter of each "doubled" polygon from the perimeter of the one before. His final result was that π lies between 3 1/7 and 3 10/71.*

to obtain the first 35 digits of π. (See "Figure "Tombstone" on p. 28.) Incidentally, this is more than enough accuracy for any conceivable practical purpose. Even if a circle was perfect down to the atomic scale, the thermal vibrations of the molecules of ink would make most of those digits physically meaningless. Thus, for the last four centuries (and arguably since Archimedes), mathematicians' attempts to compute π to ever greater precision have been driven primarily by curiosity about the number itself. In recent years, a second motivation has been the desire to improve the techniques of high-precision computing, for which computations of π have become a benchmark.

The seventeenth century brought a radically improved method of computing π, using the tools of calculus rather than geometry. Mathematicians discovered numerous infinite series associated with π. The most famous is known as Leibniz's formula,

$$\frac{\pi}{4} = 1 - \frac{1}{3} + \frac{1}{5} - \frac{1}{7} + \frac{1}{9} - \frac{1}{11} + \cdots.$$

Although it is one of the most striking formulas in calculus, Leibniz's formula is completely impractical for approximating π to more than a couple of decimal places because it converges very slowly. Newton, who along with Leibniz was one of the co-discoverers of calculus, found a more efficient formula based on an arcsine integral:

$$\pi = \frac{3\sqrt{3}}{4} + 24\left(\frac{1}{3 \cdot 8} - \frac{1}{5 \cdot 32} - \frac{1}{7 \cdot 128} - \frac{1}{9 \cdot 512} - \cdots\right).$$

The first term to the right of the parentheses comes from finding the area of the blue triangle in Figure 2; the remaining terms come from setting up an integral to evaluate the area of the red region on the left and evaluating it with an infinite series. In this way, Newton computed π to 15 digits. (In a famous letter to a friend, written in 1666, Newton confessed "I am ashamed to tell you to how many figures I carried these computations, having no other business at the time.")

Most of the infinite series for π are based on an infinite series for the inverse tangent function, $\arctan x = x - \frac{1}{3}x^3 + \frac{1}{5}x^5 - \frac{1}{7}x^7 + \cdots$ and the fact that $\tan(\pi/4) = 1$ (i.e., $\arctan(1) = \pi/4$). Leibniz's formula simply lets $x = 1$. To get more rapid convergence, it is necessary to use smaller values of x. One pretty method uses the geometric diagram in Figure 3 (see next page), from which it is easy to show that

$$\pi = 4\left(\arctan\left(\frac{1}{2}\right) + \arctan\left(\frac{1}{3}\right)\right).$$

[To see that this is true, notice that $\angle CDH = \pi/4$ and that $\angle CDE = \arctan(\frac{1}{2})$, from the two corresponding right triangles. Meanwhile, $\angle GDH = \arctan(GH/DH) = \arctan(GH/CH) = \arctan(1/3)$, because of the classic result from Euclidean geometry that the medians of a triangle (in this case $\triangle ACD$) divide each other in a 1:3 ratio.] Starting with the above trigonometric identity, which was discovered by Euler in 1738, one can substitute $\frac{1}{2}$ and $\frac{1}{3}$ for x into the arctangent series. Even the first five terms of the series (or ten terms

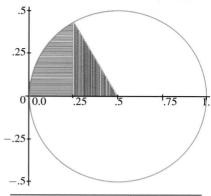

Figure 2. *The geometric reasoning behind Newtons calculus-based 15-digit approximation of pi. The circular sector shown has area $\pi/24$. The area of the blue triangle is easily computed by elementary geometry. The area of the red triangle can be computed by an integral, which Newton rewrote as an infinite series. By stopping the summation after a finite number of terms, one can compute any desired number of decimal digits of π (with enough patience). (Figure courtesy of Jonathan Borwein.)*

total—five for $x = \frac{1}{2}$ and five for $\frac{x-1}{3}$) are enough to obtain a pretty good approximation, $\pi \approx 3.1417\ldots$.

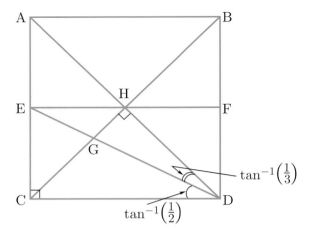

Figure 3. *Another geometrically inspired formula for pi, discovered by Euler:* $\pi = \arctan 1 + \arctan 2 + \arctan 3$, *is equivalent to* $\pi/4 = \arctan \frac{1}{2} + \arctan \frac{1}{3}$. *Angle CDG has measure* $\arctan \frac{1}{2}$, *and angle GDH has measure* $\arctan \frac{1}{3}$. *The arctangents can be computed by using the well-known convergent series for* $\arctan x$ *from calculus. (Based on a Figure by Michael Ecker, College Mathematics Journal, May 2006, p. 219. Courtesy of the Mathematical Association of America.)*

Even better is a formula discovered by the English mathematician John Machin and first published in 1706:

$$\frac{\pi}{4} = 4\arctan\frac{1}{5} - \arctan\frac{1}{239}.$$

Getting two digits of π requires only three terms: $\pi \approx 16(\frac{1}{5} - \frac{1}{375}) - \frac{4}{239} \approx 3.14$. Machin's formula became the basis for almost all computations of π for the next two centuries. Machin himself computed a hundred digits, and the zenith of his formula's popularity undoubtedly arrived in 1853, when William Shanks published a book in which he calculated 607 digits of π. Twenty years later, he added another hundred digits, giving π to 707 digits. (Shanks also calculated extensive tables of prime numbers, reciprocals, and digits of another famous number known as Euler's constant.)

Shanks's calculation stood unchallenged until 1945, when D.F. Ferguson, working with a different formula ($\pi = 12\arctan(1/4) + 4\arctan(1/20) + 4\arctan(1/1985)$), produced a new approximation—and discovered an error in Shanks's value. Not only that, Ferguson managed to find where Shanks made his mistake. Shanks had omitted a zero in the 531st decimal place of $1/497 \cdot 5^{497}$, so that the number $\ldots 80482897\ldots$ was incorrectly replaced by $\ldots 8482897\ldots$. As a consequence (after multiplying the $\arctan(1/5)$ by 16) Shanks's value of π was correct to only 527 digits.

Ferguson went on to calculate 710 digits of π, using a mechanical calculator. This was a first, as all earlier approximations had been done by hand, but an even bigger change was coming. By 1949, ENIAC, one of the first electronic computers, set a new record of more than 2000 digits. As electronic computers improved, the number of known digits of π took off into the stratosphere: ten thousand digits in 1958, a hundred thousand in 1961, and a million in 1973.

Remarkably, one of the obstacles to high-precision computation of π, which surfaced as early as the 1950s, was the inefficiency of the "grade school" method of multiplying two numbers together. In many-digit computations, computer programmers typically store 32 bits (or 8 decimal digits) in each computer "word." Computing the product of two N-digit numbers by the conventional method involves multiplying every pair of computer words, so that a product of two N-digit numbers requires a constant times N^2 multiplications, plus a comparable number of additions. In the 1960s, computer scientists realized that there is a much faster method for large numbers, which involves first taking a fast Fourier transform (FFT) of each of the strings of words. This elegant scheme simplifies the computation so that it requires only a constant times $N \log N$ steps, which for large N is much less than N^2.

Unfortunately, FFTs turn out to be difficult on today's massively parallel supercomputers because of the need for massive exchanges of data between nodes of the system. It is better, if possible, to structure the computation so that FFTs are only performed on a single node, and other schemes are used between nodes. In short, a state-of-the-art computation of π turns out to be even more difficult than it appears at first glance; the computer programmer has to be very careful about how to implement the calculations efficiently. One might even say that the mathematics is the easiest part!

Nevertheless, Kanada and colleagues, who have set most of the records since the early 1980s, reached the billionth digit of π by 1989. And in 2002, they pushed the record to a staggering one trillion digits. The trillion-digit computation took some 600 hours. It was performed by a 64-processor parallel supercomputer built by Hitachi. Kanada's team actually computed π twice, using two different formulas (see Figure 4 next page). The calculations were initially carried out in hexadecimal, or base-16, which is convenient for computers but also has theoretical significance (as explained below). In hexadecimal, the "digits" for ten through fifteen are usually denoted by the letters A through F, so that the hexadecimal expansion of π is 3.243F6A8..., which means $3 + (2/16) + (4/16^2) + (3/16^3) + (15/16^4) + \cdots$. The computer calculated just over a trillion hexadecimal digits of π with each formula, and checked that the results agreed. It then ran an algorithm to convert the hexadecimal result into an ordinary, base-10 expansion. Because base 10 is smaller than base 16, the conversion produced 1.2411 trillion decimal digits. As one final check, Kanada's team converted their decimal expansion *back* into hexadecimal and verified that it agreed with the original result.

> **The trillion-digit computation took some 600 hours. It was performed by a 64-processor parallel supercomputer built by Hitachi.**

In December 2002, Kanada computed π to over **1.24 trillion decimal digits**. His team first computed π in hexadecimal (base 16) to 1,030,700,000,000 places, using the following two arctangent relations:

$$\pi = 48\tan^{-1}\frac{1}{49} + 128\tan^{-1}\frac{1}{57} - 20\tan^{-1}\frac{1}{239}$$
$$+48\tan^{-1}\frac{1}{110443}$$

$$\pi = 176\tan^{-1}\frac{1}{57} + 28\tan^{-1}\frac{1}{239} - 48\tan^{-1}\frac{1}{682}$$
$$+96\tan^{-1}\frac{1}{12943}$$

due to Takano (1982) and Stöfmer (1896). Kanada verified the results of these two computations agreed, and then converted the hex digit sequence to decimal.
The resulting decimal expansion was checked by converting it back to hex.

Figure 4. *The formulas Kanada used for computing pi. (Figure courtesy of Jonathan Borwein.)*

The new calculation also matched up with the 206 billion digits of the 1999 computation. Kanada's team did one more spot check on the correctness of their computation: Using another algorithm, they recomputed 20 hexadecimal digits beginning with the trillion-and-first. The result (B4466E8D21 5388C4E014, if you must know) agreed with the main computation.

The algorithm for computing isolated digits of π is based on a remarkable formula discovered in 1996 by David Bailey, now at the Lawrence Berkeley Laboratory, Peter Borwein at Simon Fraser University in Burnaby, British Columbia, and Simon Plouffe, now at the University of Montreal. The BBP formula, as it's called, looks on the surface like many other infinite series approximations for π (see Figure 5), but it represents another major breakthrough in π computation. For the first time, a computer can compute the hexadecimal digits of π starting anywhere. That is, to obtain the trillion-and-first hexadecimal digit, it is not necessary to compute any of the first trillion digits. Until Bailey, Borwein and Plouffe discovered their formula, no one had even dreamed such a thing was possible.

The other unusual aspect of the BBP formula is the way it was discovered. Most formulas for π—indeed, most formulas of any sort—are derived by hand, starting from other known formulas. The BBP formula, though, was the product of a computer search. In 1995, Borwein and Plouffe realized that any infinite sum whose terms are of the form $P(k)/b^k Q(k)$, where P and

Q are polynomials, permits the rapid determination of individual base-b digits. Their starting point was a familiar formula for the natural logarithm of 2, $\ln 2 = \frac{1}{2} + \frac{1}{2 \cdot 2^2} + \frac{1}{3 \cdot 2^3} + \cdots$, but they were soon able to find similar formulas in the mathematical literature for many other well and lesser known numbers. However, their literature search did not turn up anything for π. They had better luck looking in the computer.

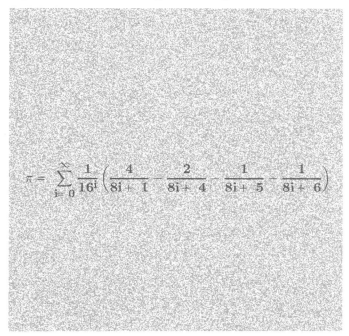

$$\pi = \sum_{i=0}^{\infty} \frac{1}{16^i}\left(\frac{4}{8i+1} - \frac{2}{8i+4} - \frac{1}{8i+5} - \frac{1}{8i+6}\right)$$

Figure 5. *The Borwein-Bailey-Plouffe formula for pi, which for the first time made it possible to compute any given digit of pi without knowing all the previous digits. However, it works only in base 16 (or more generally, a power of 2). In the background is a computer image of the first 250,000 binary digits of pi, with each 0 rendered as a dark pixel and each 1 rendered as a light pixel. The image is 500 pixels by 500 pixels. (Figure courtesy of Jonathan Borwein.)*

Borwein and Plouffe turned to an algorithm called PSLQ, developed by Bailey and Helaman Ferguson, who was then at the Supercomputing Research Center in Lanham, Maryland. PSLQ is a streamlined version of an algorithm discovered in the 1970s by Ferguson and Rodney Forcade of Brigham Young University for numerically detecting integer relations among real numbers. Borwein and Plouffe input π along with all the numbers for which they had found base-2 type infinite sums. After several weeks—they restarted the computation several times as they found more formulas in the literature—PSLQ discovered a remarkably simple relation involving the natural logarithm, the arctangent, and a special function known as the hypergeometric function, usually written $F(a,b;c;z)$. The number π, it turns out, is equal to $4F(1/4,5/4;1;-1/4) + 2arctan(1/2) - \ln(5)$. Each of these three terms can be expanded as a sum of the type Borwein and Plouffe had been studying. When the computational dust settled, they had a striking new formula for π (see Figure 6), which can be used to compute individual digits of π in base 16 (hexadecimal notation). As often happens in mathematics, after the fact they discovered a more elementary proof of the formula that uses only techniques from a first-year calculus course (and no hypergeometric function).

Incidentally, it remains impossible to compute the trillion-and-first digit in the *decimal* expansion of π without computing all the preceding ones. The BBP formula only makes this trick possible in base 16 (as well as 2, 4, and 8). There are strong grounds for believing that no such formula exists in base 10, the base humans normally use for arithmetic. It seems as if an accident of our anatomy (the fact that we have 10 fingers) has stuck us with the "wrong" notation system for computing π.

The analysis of BBP-type formulas has led Bailey and Richard Crandall, a computational number theorist at Reed College in Portland, Oregon, to some intriguing new results in the study of so-called "normal" numbers. Roughly speaking, a number is normal (base 10) when its decimal expansion doesn't favor any string of digits over any other—that is, each digit appears, on average, ten percent of the time, each *pair* of digits appears one percent of the time, and so forth. There is a similar meaning for being normal base-b: Every string of k base-b digits appears, in the limit, with frequency $1/b^k$. In other words, it seems as if the digits in base b have been generated by a random-number generator.

It is suspected that π is normal in every base. Such numbers are called "absolutely normal." Curiously, even though almost all real numbers are absolutely normal ("almost all" has a technical meaning in real analysis), there is not a single known example of such a number! In fact, researchers suspect that all of the irrational numbers they encounter in "normal" mathematics, including square roots and logarithms such as $\sqrt{2}$ or $\ln 2$, are absolutely normal, but none of these numbers is yet known to be normal in any base, much less all bases. (Rational numbers, of course, have no hope of being normal, since they are characterized as having repeating—and therefore highly non-random—decimal expansions.)

Kanada's latest calculation of π exhibits no sign of abnormality in either base 10 or base 16 (see Figure 6). But no calculation of digits can ever settle the matter one way or the other. It is always possible that π could go through a trillion digits without any sign of abnormality, and then suddenly start churning more 9's than 0's. Such behavior would seem too perverse to be credible—but nevertheless, in the absence of a solid theoretical proof it cannot be ruled out. No such theorem has been proved for π. However, Bailey and Crandall have proved a criterion for base-b normality for numbers with a BBP-type formula, and they have used their theorem to prove normality for one class of such numbers.

> Curiously, even though almost all real numbers are absolutely normal ("almost all" has a technical meaning in real analysis), there is not a single known example of such a number!

Decimal Digit	Occurrences
0	99999485134
1	99999945664
2	100000480057
3	99999787805
4	100000357857
5	99999671008
6	99999807503
7	99999818723
8	100000791469
9	99999854780
Total	**1000000000000**

Figure 6. *The number of occurrences of each decimal digit in the first trillion digits of pi. The digits appear to occur equally often, with a few random fluctuations, as one would expect if pi is a normal number. However, no one has proved yet that pi is normal, and in fact no specific instance of a normal number is known. (Figure courtesy of Jonathan Borwein.)*

Bailey and Crandall's normality condition for a number α given by a BBP-type formula $\sum P(k)/b^k Q(k)$ involves looking at a sequence of fractions between 0 and 1 that starts $x_0 = 0$ and, for $k = 1, 2$, etc., defines x_k as the fractional part of $bx_{k-1} + P(k)/Q(k)$. For $\ln 2 = \sum 1/2^k k$, the sequence thus defined begins 0, 0, 1/2, 1/3, 11/12, 1/30, 7/30, 64/105, 289/840, etc. Their theorem says that α is base-b normal if and only if the sequence of x_ks is equidistributed—that is, no subinterval of $[0, 1]$ gets more (or less) than its fair share of x_ks (more precisely, the fraction of x_ks landing in any subinterval of length ℓ tends to ℓ as k tends to infinity).

Using this theorem, Bailey and Crandall have proved normality for a particular class of numbers with BBP-type formulas: Numbers of the form $\sum 1/b^{c^k} c^k$ are base-b normal whenever b and c are relatively prime (i.e., have no prime divisor in common). This had been proved for the case $b = 2, c = 3$ by Richard Stoneham in 1970; the new proof is simultaneously simpler and more general. Unfortunately, these numbers that have been "cooked up" as examples of normality do not have any known independent meaning as roots of polynomials or values of significant functions.

Proving the normality of π, or any other familiar irrational number, therefore remains a daunting challenge. But it wasn't so long ago that computing a trillion—or even a mere billion—digits of π seemed forever out of reach. Stay tuned.

> **Proving the normality of π, or any other familiar irrational number, therefore remains a daunting challenge.**

Irrationalities

The discovery that the square root of 2 cannot be expressed as the ratio of two whole numbers was one of the turning points of ancient Greek mathematics. According to one legend, the fact was discovered by members of the Pythagorean brotherhood, who sought to keep it secret. When the secret was revealed by one of their members (the Deep Throat of his day), the brothers reacted by tossing him overboard during a sea voyage.

The irrationality of $\sqrt{2}$ is easy to prove. For centuries mathematicians wondered about π. Finally, in the eighteenth century, they found a proof that π is irrational. Unlike $\sqrt{2}$, however, the proof is complicated. Later in the nineteenth century, it was proved that π is not only irrational but "transcendental," meaning that it is not the root of any polynomial with integer coefficients. (Numbers that are roots of polynomials, such as $\sqrt{2}$, which is a root of $x^2 - 2$, are called algebraic.) In general, proofs of transcendentality are very complicated.

Many other mathematical constants are known to be irrational, and quite a few are known to be transcendental. Others have remained stubbornly mysterious. Among them are certain special numbers associated with number theory's most famous function, the Riemann zeta function, $\zeta(s) = \sum 1/n^s$ (see "A Prime Case of Chaos," *What's Happening in the Mathematical Sciences*, Volume 4). For positive integers $s = 2n$, the value of the zeta function is a rational multiple of π^{2n}. For example, $\zeta(2) = \pi^2/6$, $\zeta(4) = \pi^4/90$, $\zeta(6) = \pi^6/945$, and so forth. All this was discovered by the Swiss mathematician Leonhard Euler in the eighteenth century, who proved the irrationality of the number e.

It may seem natural to suppose that the same is true for odd values of s as well. But it is almost certainly *not* true. The ratio $\zeta(3)/\pi^3$ has been computed to millions of decimal places, with no sign of a repeating pattern. If the ratio is a fraction, the numerator and denominator are extremely large.

Because π is transcendental, its powers are all irrational, and hence the values of the zeta function at even values of s are all irrational (indeed, transcendental). Can anything be said about the irrationality of the zeta function at odd values of s? Yes. In 1979, Roger Apéry at the University of Caen in France surprised number theorists with a proof that $\zeta(3)$ is irrational. Apéry's proof, while complicated, did not use any modern methods; it was dubbed "the proof that Euler missed."

It was initially anticipated that Apéry's proof would open the floodgates for similar results on the zeta function at other odd values of s. It hasn't worked out that way. Recently, however, Tanguy Rivoal at the Université de Grenoble in France and Wadim Zudilin at Moscow State University in Russia have opened the sluices a bit.

Rivoal has shown that infinitely many of the values $\zeta(3)$, $\zeta(5)$, $\zeta(7)$, $\zeta(9)$, etc. are irrational—and not only irrational, but *independently* irrational. That is, there are infinitely many odd numbers $2n + 1$ for which $\zeta(2n + 1)$ cannot be written as a rational combination of the form $r_1 + r_3\zeta(3) + \cdots + r_{2n-1}\zeta(2n - 1)$ with rational numbers $r_1, r_3, \ldots, r_{2n-1}$. In fact, Rivoal showed that the number of independent irrationals among the numbers first n odd values of the zeta function grows at least with the logarithm of n (more precisely, it exceeds $(\ln n)/3$). By carefully examining the estimates that underlie his proof, Rivoal was able to show that at least one of the nine numbers $\zeta(5), \zeta(7), \ldots, \zeta(21)$ is irrational.

Zudilin found similar results independently, and has strengthened some of Rivoal's results. In particular, Zudilin has shown that there is an irrational number among the next four values of the zeta function, $\zeta(5)$, $\zeta(7)$, $\zeta(9)$, $\zeta(11)$. He has also shown that there is an irrational value independent of $\zeta(3)$ among the values $\zeta(5)$ up to $\zeta(145)$ (Rivoal had gotten an upper limit of 169).

Further research will undoubtedly continue to chip away at the irrational nature of the zeta function. At least today's mathematicians don't worry about getting thrown overboard for making their discoveries public.

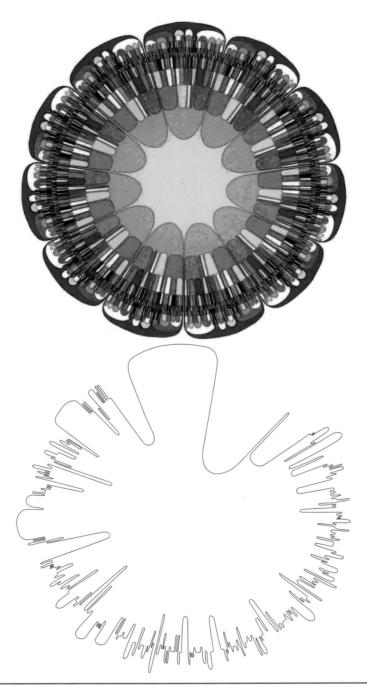

"11-set Doily." *Peter Hamburger's 11-set "doily." Top, the entire rotationally symmetric Venn diagram. Bottom, one of the 11 individual curves that makes up the diagram. (The other 10 curves are all, of course, rotated versions of this one.) (The mathematical foundation that made it possible to create the figures by artist Edit Hepp was made by Peter Hamburger; they are his intellectual properties; and he holds all the copyrights for this mathematics. The figures were created by artist Edit Hepp; they are her intellectual properties; and she holds all the copyrights.)*

Combinatoricists Solve a Venn-erable Problem

Barry Cipra

V ENN DIAGRAMS SEEM SO SIMPLE. Indeed, the topic is usually introduced in high school algebra, where overlapping circles are used to illustrate the various ways that sets can intersect (see Figure 1). Yet the subject offers a surprising number of mathematical challenges. One of these, regarding the existence of "rotationally symmetric" Venn diagrams, baffled mathematicians for over two decades, until it was recently solved—thanks in part to an inspired idea of an undergraduate math major. The detailed study of Venn diagrams falls under the heading of combinatorial geometry. Geometry because the diagrams consist of geometric shapes in the plane; combinatorial because they involve combinations of objects. Problems in combinatorics are often concerned with placing items in a list; in combinatorial geometry the problems are concerned with arranging items in space.

It helps to be precise about what a Venn diagram is. A Venn diagram for n sets consists of n closed curves. The "curves" need not be smooth—they may in fact be polygonal— and they need not all have the same size or shape (see Figure 2, next page). Each pair of curves can (and indeed must) intersect at one or more points, but not infinitely often. In particular, there are no "shared arcs" in a Venn diagram. When all drawn at once, the curves cut the plane into regions that are *inside* certain curves and *outside* the others. The crucial, defining feature of a Venn diagram is that each such inside/outside combination is represented by exactly one such region.

The number of distinct regions in a Venn diagram grows exponentially: With n sets, there are 2^n regions, including the common interior and common exterior. This is where the complications begin to creep in: As n gets larger and larger, those 2^n regions, it turns out, can be arranged in lots and lots of different ways.

This is not at all evident in the familiar 2- and 3-set pictures. Indeed, there is essentially only one way to draw a Venn diagram with two sets: No matter how wildly you draw the two curves, they can be thought of as just a badly drawn pair of circles. (This is all the more true of a 1-set Venn diagram—and totally trivial if there are no sets!) However, there are 14 differ-

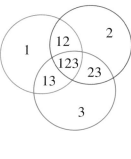

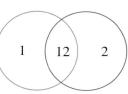

Figure 1. *The familiar 2-set and 3-set Venn diagrams using circles. A Venn diagram consists of n simple closed curves that divide the plane into 2^n connected regions, in such a way that each "inside-outside" combination is represented exactly once. For instance, if there are two curves A and B, the four combinations are "inside-inside," "inside A-outside B," "outside A-inside B," and "outside-outside." Both of these diagrams are rotationally symmetric. (Figure from "Venn Diagrams and Symmetric Chain Decompositions in the Boolean Lattice," J. Griggs, C. E. Killian, and C. D. Savage, Electronic Journal of Combinatorics 11 (2004), Research Paper 2, 30pp. electronic.)*

ent ways to draw a 3-set Venn diagram, grouped into 6 classes (see Box, "Vennis Anyone" and Figure 3, page 44).

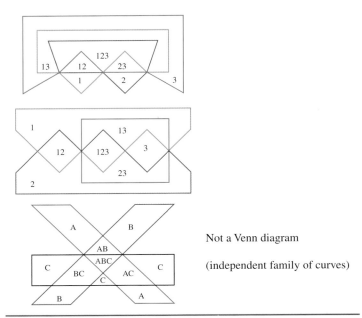

Not a Venn diagram

(independent family of curves)

Figure 2. *Venn diagrams do not have to be drawn with smooth curves. Here are two polygonal 3-set Venn diagrams, and one "non-Venn diagram." In the non-Venn diagram there is an inside-outside combination that consists of two disjoint regions; can you find it? (Figure from "Venn Diagrams and Symmetric Chain Decompositions in the Boolean Lattice," J. Griggs, C. E. Killian, and C. D. Savage, Electronic Journal of Combinatorics 11 (2004), Research Paper 2, 30pp. electronic.)*

When there are more than three sets, the enumeration problem is unsolved, except for a special type of diagram known as a *simple* Venn diagram. A Venn diagram is said to be simple if every point of intersection has exactly two curves crossing. A famous theorem known as Euler's formula, applied to simple Venn diagrams, shows that there are $2^n - 2$ curve crossings in a simple n-set Venn diagram (see Box, "Venn Meets Euler," p. 45). For three sets, there is just one simple Venn diagram, as Figure 3 shows. For four sets, there are two simple Venn diagrams, both belonging to the same class. For $n = 5$, the number of classes of simple Venn diagrams jumps to 19. Beyond five, the number of classes has not been tallied.

In 1963, David Henderson at Swarthmore College took note of another property that is obviously true for the familiar 2- and 3-set Venn diagrams, but less obvious—and, indeed, *not* true—in general: The familiar diagrams are "rotationally symmetric." That is, there is a point in the region corresponding to the common intersection ($A \cap B$ for $n = 2$, and $A \cap B \cap C$ for $n = 3$) about which the diagram is symmetric under rotation by $360/n$ degrees (or $2\pi/n$ radians). More precisely, a Venn diagram is rotationally symmetric if it can be drawn by drawing one close curve and then rotating it by multiples of $2\pi/n$ radians about a point in its interior to produce the other curves. Are there such diagrams, Henderson asked, for values of n greater than three?

Vennis, Anyone?

Venn diagrams are traditionally drawn on paper. Maybe they should really be drawn on tennis balls.

The plane and the sphere are closely connected. In particular, the sphere can be thought of as the plane with an extra "point at infinity." (Alternatively, the plane is a "punctured" sphere.) The connection comes courtesy of a correspondence called stereographic projection: Place a transparent globe so that its south pole sits on an (infinite) tabletop, and shine a laser pointer from its north pole to a point on the table. (See "Stereographic Projection," p. 51.) On its way, the laser beam passes through a point on the globe. That is the unique point on the sphere that corresponds to the chosen point on the table, and vice versa. As the point on the table moves further and further away from where the globe is sitting, its corresponding point on the sphere moves closer and closer to the north pole. Thus the north pole itself corresponds to the "point at infinity" for the plane.

If you draw a Venn diagram on a sphere, say using magic markers on a tennis ball, the curves separate the sphere into 2^n regions (n being the number of curves). But there is no longer an obvious "inside" and "outside" for each curve; instead, there is just "the side that contains the north pole" and "the side that doesn't." In a planar Venn diagram, there is one region that appears much different from the others— the unbounded region, which surrounds the whole diagram. But on a sphere, there is no such distinction; the region that contains the north pole looks pretty much like all the others.

This is actually an advantage because one is free to take the sphere and reposition it so that some other point, inside one of the other regions, becomes the north pole, and then re-project onto the plane. The result is another Venn diagram. It may, of course, be equivalent to the first, but it may not. When $n = 4$, for example, there are regions bounded by three curves and regions bounded by four, so there are two distinct projections, depending on which type of region is chosen to contain the north pole. The various diagrams, while "planely" different, are said to belong to the same class. That is why, in the enumeration problem, the number of *classes* of Venn diagrams with n sets is smaller than the absolute number of Venn diagrams.

The answer—or part of it—is "not if n is composite." The reason, as Henderson showed, is fairly simple. Aside from the common interior and common exterior of the n curves, each region rotates through a total of n copies of itself, each of the same "type," meaning the number of sets whose intersection is that region. This means that n must divide the number of pairwise intersections, the number of three-way intersections, etc., on up to the number of $n - 1$-way intersections. These numbers are well known as the binomial coefficients $\binom{n}{k} = n!/k!(n-k)!$

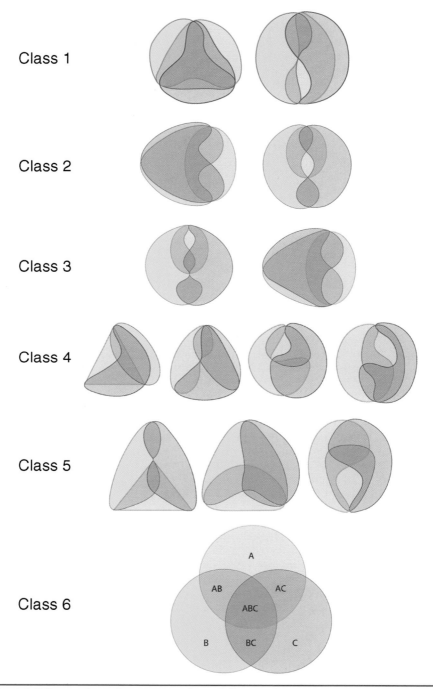

Class 1

Class 2

Class 3

Class 4

Class 5

Class 6

Figure 3. *The 14 different 3-set diagrams. A "simple" Venn diagram has no threefold crossings; note that only one of the 14 3-set diagrams is simple. (Based on original figures supplied by Frank Ruskey at the University of Victoria.)*

for $k = 2, 3$, up to $n - 1$ (see Figure 4). But a famous theorem, originally proved by Leibniz, says that n divides all its binomial coefficients if and only if n is prime. For example, $n = 4$ does not divide $\binom{4}{2} = 6$. So the six regions that correspond to pairwise intersections cannot be arranged around a central point with fourfold symmetry.

Venn Meets Euler

The great Swiss mathematician Leonhard Euler was actually one of the first mathematicians to study Venn diagrams—more than 100 years before John Venn took up the subject! Euler also studied more general types of *planar graphs*, which consist of curves, called edges; connecting points, called vertices; separating the plane into distinct regions, called faces. (A graph is called planar when none of its edges cross.) Euler showed that, for any planar graph, the numbers of vertices V, edges E, and faces F satisfy the formula $V - E + F = 2$.

For Venn diagrams, the number of faces is 2^n. For simple Venn diagrams, it is easy to see that there are twice as many edges as vertices, i.e., $E = 2V$ (see Figure 1, p. 41). Thus Euler's formula implies $V - 2V + 2^n = 2$, or $V = 2^n - 2$.

Leibniz's theorem leaves open the possibility that prime values of n do permit rotationally symmetric Venn diagrams. Henderson gave two examples for $n = 5$, one with irregular pentagons and one with quadrilaterals. He also claimed to have an example for $n = 7$ with irregular hexagons, but was unable to reproduce it later.

Branko Grünbaum at the University of Washington took up the problem in 1975. He gave additional examples for $n = 5$, including one using ellipses (see Figure 5). Frank Ruskey at the University of Victoria later did an exhaustive computer search for symmetric Venn diagrams with five sets. He found there are 243 different examples, of which only one—Grünbaum's ellipses—is simple.

Grünbaum returned to the problem in 1992, when he discovered a 7-set example (see Figure 6, page 46). Shortly after, additional 7-set diagrams were found by Anthony Edwards at Cambridge University and, independently, by Peter Winkler at Dartmouth College and Carla Savage at North Carolina State University. With these examples, it became natural to conjecture that rotationally symmetric Venn diagrams with n sets exist for *all* primes n. But, there, things came to an apparent impasse: Venn diagrams with 5 or 7 sets are still small enough that a combination of cleverness and patience suffices to ferret out examples of symmetry, but with 11 sets the number of possibilities to pick through becomes intractably large. As Grünbaum put it in a status report in 1999, "The sheer size of the problem for 11 curves puts it beyond the reach of the available approaches through exhaustive computer searches."

Enter Peter Hamburger. A graph theorist at the University of Indiana–Purdue University at Fort Wayne, Hamburger had worked on some of the enumeration problems for Venn diagrams. (In a series of papers with Raymond Pippert and Kiran

Pascal's Triangle

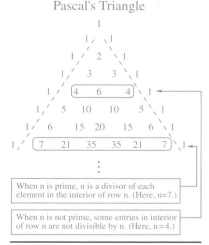

When n is prime, n is a divisor of each element in the interior of row n. (Here, n=7.)

When n is not prime, some entries in interior of row n are not divisible by n. (Here, n=4.)

Figure 4. *Pascal's triangle can be used to prove that there are no rotationally symmetric Venn diagrams with 4 sets. The k-th entry in the n-th row of Pascal's triangle (counting the 1 at the top as the zero-th row) enumerates how many regions are inside k curves but outside $(n - k)$ curves. If the Venn diagram is rotationally symmetric, each of the entries (not including the 1s at the beginning and end) must be divisible by n. But the middle entry of the fourth row is not divisible by 4. In fact, for any non-prime n, the nth row of Pascal's triangle contains some elements not divisible by n, hence a Venn diagram with n-fold rotational symmetry is impossible.*

Figure 5. *Branko Grünbaum's rotationally symmetric Venn diagram with 5 regions, all bounded by ellipses. (Figure courtesy of Branko Grünbaum.)*

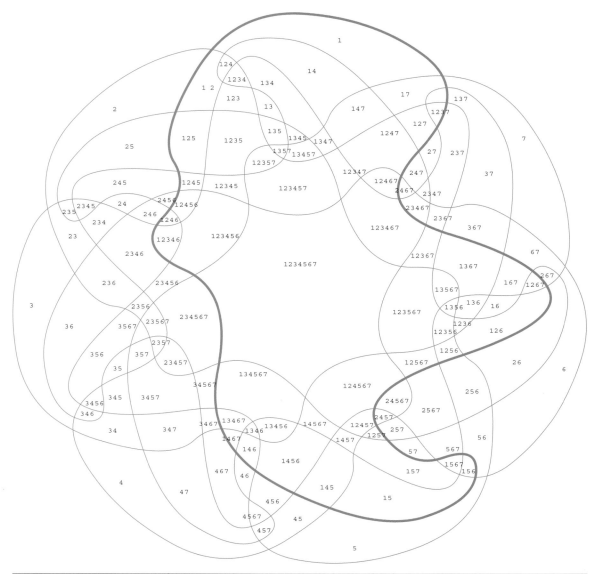

Figure 6. *Grünbaum's rotationally symmetric Venn diagram with 7 regions. (Figure courtesy of Branko Grünbaum.)*

Chilakamarri, he identified the 14 different Venn diagrams with three sets and the 19 classes of simple Venn diagrams with five sets.) Hamburger thought there might be a way to construct rotationally symmetric diagrams using methods developed for a different, purely combinatorial problem: symmetric chain decompositions of Boolean lattices.

A Boolean lattice on n elements can be viewed as the subsets of the set $\{1, 2, \ldots, n\}$, stratified by the size of the subsets, and with each subset "connected" to the subsets one level above and below it that differ by the additional presence or absence of one element (see Figure 7a). A "chain" is simply a string of connected subsets, and a "chain decomposition" is a collection of disjoint chains that account for all the subsets. (Some of the "chains" may consist of a single subset. In fact, letting each subset constitute its own chain is a perfectly good chain decomposition.) A chain decomposition is "symmetric" if the smallest and largest subset in each chain are of complementary size— that is, if the smallest subset in a chain has k elements, then the largest subset in that chain has $n - k$ elements (see Figure 7b).

(a)

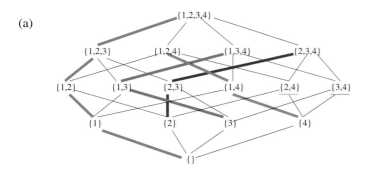

(b)

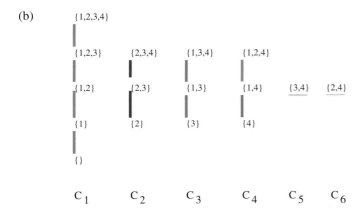

Figure 7. *A chain decomposition of a Boolean lattice is a partition into several "chains," in which each chain consists of a collection of sets that differ by turning one "inside" into and "outside" or vice versa. It is trivial to draw chain decompositions if you are given a Venn diagram. The key step in constructing rotationally symmetric Venn diagrams is to do the reverse: to go from an abstract chain decomposition with the right kind of symmetry properties to constructing a concrete Venn diagram. Peter Hamburger showed how to do this for an 11-set Venn diagram, and Carla Savage, Peter Griggs, and Chip Killian accomplished it for all remaining primes. (Figure from "Venn Diagrams and Symmetric Chain Decompositions in the Boolean Lattice," J. Griggs, C. E. Killian, and C. D. Savage, Electronic Journal of Combinatorics **11** (2004), Research Paper 2, 30pp. electronic.)*

Starting with a Venn diagram, it is easy to draw chains by connecting adjacent regions in the diagram (always crossing curves from inside to out), and therefore easy to create chain decompositions. If the diagram is rotationally symmetric, the chains (ignoring the full and empty sets) can be chosen so as to rotate onto one another. The result is not guaranteed to be a symmetric chain decomposition. However, sometimes it is.

Hamburger's idea was to go the other way: Start with a symmetric chain decomposition and systematically build a Venn diagram around it. Moreover, by imposing an additional rotational symmetry on the chains, the resulting diagram would be rotationally symmetric.

There was no guarantee this approach would work, but the two symmetry conditions and the systematic construction sharply reduced the number of possibilities to be considered. Examples for $n = 5$ and 7 turn out to be easy to find. Hamburger then set his sights on the next, unsolved case: $n = 11$.

The method worked. Hamburger was able to construct an 11-set "Doily"—his term for the kind of Venn diagram that comes from his method, because of their lacy look—that is rotationally symmetric (see "11-set Doily," p. 40). His wife, Edit Hepp, who is an artist, has turned several of his examples into works of art. Hepp uses colored pencils to fill in regions corresponding to different types of intersections. The results are beautiful, mandala-like abstracts.

Hamburger's approach opened a new route for constructing rotationally symmetric Venn diagrams, but it still required cleverness and patience in picking through possibilities— enough of them that the next case, $n = 13$, was too dense a thicket for Hamburger to penetrate. It ultimately took an undergraduate to find a way.

Jerry Griggs at the University of South Carolina read Hamburger's paper and started thinking about how to obtain Venn diagrams from symmetric chain decompositions. He found that the construction depended on what he called a chain-covering property, which he showed could always be satisfied, even for composite n. To get rotational symmetry for prime values of n, it would be necessary (and sufficient) to impose rotational symmetry on the chain decomposition while maintaining the chain-covering property.

Imposing symmetry amounts to deciding which regions occur in a single "wedge" of the diagram and then requiring the next wedge contain the same regions "moved up" by one. For example, if one wedge contains the region corresponding to the interiors of curves 1, 3, and 6 (and the exteriors of curves 2, 4, 5, and 7, for $n = 7$), then the next wedge must contain the region corresponding to the interiors of curves 2, 4, and 7, the wedge after that must contain the region corresponding to the interiors of curves 3, 5, and 1, and so forth.

A nice way to abbreviate all this is to use binary strings to describe the regions: 1010010 is in one wedge, 0101001 in the next, 1010100 in the one after that, and so forth. When n is prime, every such string, except for all 1's or all 0's (corresponding to the center and the unbounded, outer region of the Venn diagram), belongs to an "orbit" of n strings. For a diagram to be symmetric, each wedge must include exactly one "representative" of each orbit. Consequently, Griggs saw, the key to rotational symmetry was to find a rule for picking these representatives in a way that preserved the crucial chain-covering property.

Like Hamburger, Griggs got stuck. "None of the rules I tried for picking representatives worked out," he says. There was another problem as well, he adds: "I never had much time to work on it."

Griggs described his ideas to Carla Savage at North Carolina State University, and Savage enlisted an undergraduate, Charles "Chip" Killian (currently a graduate student at Duke University), to work on the next case, $n = 13$. "I thought we could hack out 13," she says. It seemed like a good project for a student.

Killian hacked out a lot more than 13. He found a rule for picking representatives that worked for *all* primes.

Killian's rule is surprisingly simple. It is based on what is called the "block" structure of the binary string abbreviations for the regions. For example, to determine which string in the cycle generated by 01110011010 (for $n = 11$) belongs to the wedge whose outermost bound is an arc of curve number one, consider the six rotations of it that begin with a 1: 11100110100, 11001101001, 10011010011, 11010011100, 10100111001, and 10011100110. (Everything in this wedge is in the interior of the first curve, which is why it's not necessary to consider strings beginning with a 0.) In each of these candidate strings there are alternating blocks of 1's and 0's. Killian's rule is to pick the one with the least total number of 1's and 0's in the first pair of blocks—in this case, 10100111001, for which the total is 2. If there is a tie, the rule is to look at the next pair of blocks, and so forth, until the tie is broken, which is guaranteed to happen eventually.

When Killian first showed Savage his rule, she was skeptical. "It took a while for him to convince me we should take this seriously," Savage recalls. It isn't obvious the rule is copacetic with the chain-covering property. (It's also not obvious the tie-breaking procedure always works, though that's fairly easy to prove.) But within a few months, Griggs, Savage, and Killian had checked that everything worked. "There are a lot of things to verify, but once you see them they're easy to prove," Savage says.

It's somehow fitting that the last step in solving Henderson's question about rotationally symmetric Venn diagrams should be taken by a student: Henderson wrote his paper when *he* was an undergraduate. However, Hamburger's doilies and the general solution by Griggs, Savage, and Killian do not settle all the questions that can be asked about these diagrams. In particular, the constructions based on symmetric chain decompositions are highly non-simple: There are points, in fact, where all n curves cross. (These points define the wedges.) In general, the constructions based on symmetric chain decompositions give rise to diagrams with $\binom{n}{(n-1)/2}$ vertices, which is much smaller than the $2^n - 2$ vertices in a simple Venn diagram for n curves.

Hamburger and colleagues György Petruska at Indiana–Purdue University Fort Wayne and Attila Sali at the Alfréd Rényi Institute of Mathematics in Budapest have found ways to tease apart many of the multi-curve crossings in the $n = 11$ case. Their best result to date has 1837 vertices—167 in each wedge—or 209 shy of the 2046 vertices an 11-curve Venn diagram needs to be simple. (Hamburger's original 11-set Doily

It's somehow fitting that the last step in solving Henderson's question about rotationally symmetric Venn diagrams should be taken by a student: Henderson wrote his paper when *he* was an undergraduate.

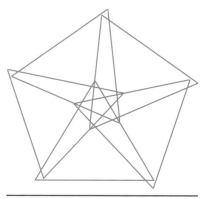

Figure 8. *Grünbaums Venn diagram with five equilateral triangles. (Figure courtesy of Branko Grünbaum.)*

has $\binom{11}{5}$ = 462 vertices.) Together with Frank Ruskey and Mark Weston at the University of Victoria, Savage and Killian have shown it's possible to increase the number of crossings in their general construction to at least 2^{n-1}, but that's still barely halfway to the $2^n - 2$ crossings for a simple Venn diagram. To date there are no known examples of a simple rotationally symmetric Venn diagram when n is greater than 7.

Venn diagrams offer many other challenges. In 1984, Peter Winkler, who is now at Dartmouth College, conjectured that every simple Venn diagram with n curves can be extended to a simple Venn diagram with $n + 1$ curves by the addition of one more curve. (Grünbaum noted that extendability had not been proved even if simplicity were not assumed. This was settled in 1996 by Hamburger, Pippert, and Chilakamarri, who showed that every Venn diagram with n curves can be extended to one with $n + 1$ curves. Their proof, however, produces non-simple extensions.)

Also in 1984, Grünbaum asked whether it is possible to draw a Venn diagram with six triangles. Grünbaum and Winkler had solved the corresponding problem for $n = 5$. There is, in fact, a simple, rotationally symmetric Venn diagram with five equilateral triangles (see Figure 8). This was settled in 1999 by Jeremy Carroll, a research scientist at Hewlett-Packard Laboratories in Bristol, England. By means of an exhaustive -computer search, Carroll found there are 126 different Venn diagrams with six triangles. His method, however, does not answer a related question: Can any of these diagrams be drawn using six equilateral triangles? Problems like these are likely to keep researchers—and their students—busy for years to come.

A Venn Diagram Roundup

Venn diagrams have so many interesting properties—in addition to rotational symmetry, simplicity, and extendability, there is monotonicity, convexity, rigidity, reducibility, and on and on—it would take a complicated, multi-curve Venn diagram to keep track of them all. A good place to start for all your Venn diagram needs is the online "Survey of Venn Diagrams," by Frank Ruskey at the University of Victoria. Originally published in 1997 in the Electronic Journal of Combinatorics (which is also where the Griggs–Savage–Killian paper appears), Ruskey's survey (www.combinatorics.org/Surveys/ds5/VennEJC.html) has been updated to include many of the latest results.

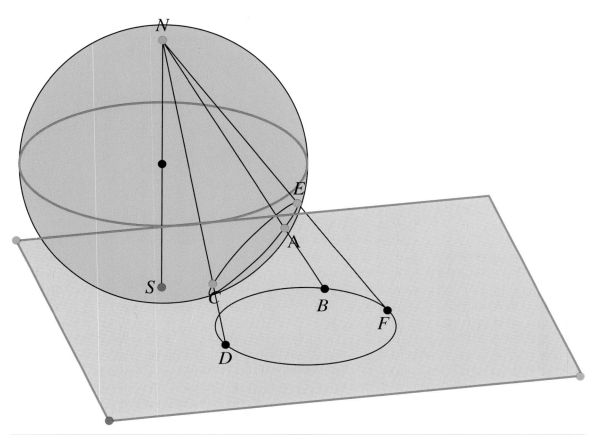

Stereographic Projection. *This figure illustrates a standard method for building a correspondence between the points on the sphere (minus the point at the north pole) and all the points on the plane. The three points labeled A, C and E on the sphere are projected onto the points B, D and F, respectively, on the (infinite) table. Likewise, the circle passing through A, C, and E projects down to the circle through B, D, and F. If the latter circle were part of a Venn diagram drawn on the table, it could be lifted back up to become part of a corresponding Venn diagram drawn on the sphere.*

Paul Erdős. *The late Hungarian mathematician was famous for offering monetary prizes for the solution of various conjectures. He offered a $3000 prize for a general theorem that would have implied that there exist arbitrarily long arithmetic progressions of primes (APPs). Tao and Green proved the latter theorem, but Erdős's prize problem remains unsolved. (Photo by George Csicsery from his documentary "N is a Number : A Portrait of Paul Erdős" (1993) © All Rights Reserved.)*

New Insights into Prime Numbers

Barry Cipra

THE FIRST FEW YEARS of the twenty-first century have seen three huge strides in the theory of prime numbers. Make that two-and-a-half huge strides. But the half stride proved instrumental for the third, which was so huge it might as well count as a stride-and-a-half.

In August 2002, Manindra Agrawal at the Indian Institute of Technology in Kanpur and two of his students, Neeraj Kayal and Nitin Saxena, announced a proof that primality can be checked by an efficient, "polynomial time" algorithm. Previous analyses of primality-checking algorithms had left open the possibility of worst-case exponential behavior or had relied on randomness.

Seven months later, Dan Goldston at San Jose State University and Cem Yildirim at Bogazici University in Turkey announced a result that dramatically improved number theorists' understanding of the gaps between primes. Their announcement, however, proved premature. Other researchers poring over Goldston and Yildirim's proof discovered an error in their reasoning. Two years and an additional coauthor later, they were back with a proof that withstood the experts' scrutiny. A large part of their original work was sound to begin with, and in fact that part had already paid off with an unexpected application to another problem.

In April 2004, Ben Green at the Pacific Institute for Mathematical Sciences in Vancouver, British Columbia, and Terence Tao at the University of California at Los Angeles announced the solution of a famous open problem in number theory: The prime numbers contain arithmetic progressions of arbitrarily long length. Green and Tao use the unflawed part of Goldston and Yildirim's original paper at a crucial stage of their proof.

To Be or Not to Be? (Prime, That Is)

Prime numbers, often called the building blocks of number theory, have long fascinated mathematicians. Euclid gave the first known proof that there are infinitely many primes, by arguing that if $p_1, p_2, \ldots, p_n$ is a list of prime numbers, then $p_1 p_2 \cdots p_n + 1$ is divisible by a prime number not on the list. Hence no finite list of prime numbers can be complete. In the eighteenth century, Euler gave a stronger proof, showing that the sum of the reciprocals of the primes, $\frac{1}{2} + \frac{1}{3} + \frac{1}{5} + \frac{1}{7} + \frac{1}{11} + \cdots$, diverges—that is, sums to infinity—which would not happen if there were only finitely many primes. While Euclid's proof is completely "elementary," in the sense that it uses only ordinary arithmetic, Euler's approach introduces the more

> Prime numbers, often called the building blocks of number theory, have long fascinated mathematicians.

sophisticated tools of mathematical analysis (such as infinite sums and limits).

Except for numbers divisible by 2 or 5, for which a quick look at the last digit is enough to identify them as composite, it takes at least a little bit of work to tell whether a number N is prime or not, and the amount of work increases with the number of digits in the number. The obvious algorithm is trial-and-error division: check for a remainder after division by the primes 3, 7, 11, 13, etc. If no prime divisor turns up that is smaller than the square root of N, then N is prime (since any exact divisor greater than $\sqrt{N}$ would have to be matched by one less than $\sqrt{N}$).

If N is composite with a small prime divisor—say less than a million—this is a reasonably efficient task for a computer to carry out. But if N is a 100-digit prime, trial-and-error would take effectively forever. It is what's known as an exponential-time algorithm, because the time it takes (in the worst case, which is when the number N is prime or has large prime divisors) grows exponentially with the number of digits in N.

Number theorists have long wondered if primality checking could be done in polynomial time—that is, by an algorithm whose run time is bounded by a power of the number of digits in the number being checked. In the theory of computational complexity, polynomial-time algorithms are considered efficient, and problems that can be solved by a polynomial-time algorithm are considered "easy" to solve. (See Box, "Time Out.")

As far as anyone knows (unless the National Security Agency is holding out on us), there is no polynomial-time algorithm for factoring composite numbers. The absence of such an algorithm is what makes the popular RSA public-key cryptosystem secure. Number theorists have made significant advances over trial-and-error, especially since the 1970s, so that factoring hundred-digit numbers is now routine (see "The Secret Life of Large Numbers," *What's Happening in the Mathematical Sciences*, Volume 3). But even the best factoring algorithms still grow at an exponential rate. Two-hundred-digit numbers are currently still safe, and thousand-digit numbers are likely to remain unfactorable, in general, for decades, barring a major breakthrough in the field.

But asking whether a number *has* prime factors is different from asking what those factors are. Number theorists have long known of some simple tests that can reveal the presence of factors without revealing the factors themselves. The simplest such test is based on a result known as Fermat's little theorem (not to be confused with the famous "last theorem"). Fermat's little theorem says that if a number N is prime, then, for any number a, N divides $a^N - a$. A mildly stronger version says that if N is prime (greater than 2) and $0 < a < N$, then, when $a^{(N-1)/2}$ is divided by N, the remainder is either 1 or $N - 1$. (In number theorese, this is written $a^{(N-1)/2} \equiv \pm 1 \mod N$.)

Time Out

Everyone who works with computers has experienced the frustration of using a program that takes too long to run. But what does "too long" mean, exactly? The same algorithm might run at different speeds on different computers; also, different users may not have the same amount of patience.

For these reasons, mathematicians have developed a way of distinguishing "fast" algorithms from "slow" ones that is independent of the hardware, the operating system, or the personal tastes of the user. Given a certain number of bits of input data, N, they work out the total number of mathematical operations the program takes to produce an answer. The best kind of program would have a run time that increases linearly with N: if you give it twice as much data, it takes twice as long to finish. As it turns out, this is too optimistic for all but the simplest kinds of problems. However, it is almost as good to have a program that runs in quadratic time (if you give it twice as much data, it takes four times as long) or cubic time (if you give it twice as much data, it takes eight times as long). All of these are considered polynomial-time algorithms because the number of operations is bounded by a polynomial N^d, where d is some fixed number.

By contrast, the run time of an exponential algorithm is bounded by a function of the form a^N, for some fixed number a. This seemingly trivial modification has huge practical consequences. If $N = 100$, for example, a quadratic algorithm might take 100^2 or 10,000 steps to finish. An exponential algorithm would take 2^{100} steps to finish the same task, a number roughly equal to 1 followed by thirty zeroes. Even on a computer that performs a trillion operations per second (comparable to today's best computers), such a program would take longer than the age of the universe to finish. Faster hardware or clever programming cannot cure the second algorithm's fundamental inability to compete. At best, all they can do is push the inevitable slowdown out to larger values of N.

These kinds of estimates of an algorithm's efficiency as N approaches ∞, called *asymptotic* estimates, are important for theory but they do not always measure the algorithm's practical usefulness. For example, the AKS primality test, described in the article, is not yet competitive with the best previously known primality tests for 100-digit numbers (the size of numbers that are currently used in coding applications). Nevertheless, we can guarantee that at some point (1000-digit numbers? 10,000-digit numbers?) the AKS algorithm will leave its competitors in the dust.

Fermat's little theorem is a test for primality to the extent that if, for some a, N does *not* divide $a^N - a$ (or, more sensitively, $a^{(N-1)/2} \not\equiv \pm 1 \mod N$ for some a between 0 and N), then N is most certainly *not* a prime. For example, 15 is not a prime because $2^7 = 128 \equiv 8 \mod 15$. Note that this calculation does not produce either factor of the number 15. For such small numbers, of course, simple trial and error (or just knowing your times tables!) is easier than computing powers and

So although it works efficiently most of the time at exposing composite numbers, Fermat's little theorem is not an all-purpose, polynomial-time, prime-proving test.

taking remainders, but when the numbers get large—say into a couple dozen digits—tests based on Fermat's little theorem and its associates become worth their weight in gold.

Why isn't Fermat's little theorem the end of the primality-testing story? For a very simple reason: It only gives a definite answer ("not prime") when N does not divide $a^N - a$; when N *does* divide $a^N - a$, the test is inconclusive. In that case N *may* be prime, but it may not—it could be a composite number "masquerading" as a prime. The first instance of this happens when $N = 341$ and $a = 2$: 341 does divide $2^{341} - 2$ (see Figure 1), but it is nonetheless composite ($341 = 31 \times 11$).

$$2^4 = 16$$
$$2^8 = 2^{4 \cdot 2} = 16^2 = 256$$
$$2^{16} = 256^2 = 65536 \equiv 64 \pmod{341}$$
$$2^{20} = 2^{16} \cdot 2^4 \equiv 64 \cdot 16 \equiv 1024 \equiv 1 \pmod{341}$$
$$2^{341} = 2 \cdot 2^{340} = 2(2^{20})^{17} \equiv 2 \cdot (1)^{17} \equiv 2 \pmod{341}$$

Figure 1. *The number 341 is a pseudoprime to the base 2. This means that 2^{341} is congruent to 2 modulo 341, yet 341 is not a prime number (because $341 = 31 \times 11$).*

For each a there are infinitely many composite numbers that "pass" the primality test based on Fermat's little theorem. In fact, there are infinitely many that pass the test for *all* values of a (see "Number Theorists Uncover a Slew of Prime Impostors," *What's Happening in the Mathematical Sciences*, Volume 1). So although it works efficiently most of the time at exposing composite numbers, Fermat's little theorem is not an all-purpose, polynomial-time, prime-proving test. (See Figure 2.)

Test	Advantage	Disadvantage
Fermat Test: $a^N \equiv a \pmod{N}$	Quick <u>necessary</u> test for primality	Not <u>sufficient</u>! Some numbers, like 341, "fool" the Fermat test into thinking they are primes.
AKS test: $(x + a)^N \equiv x^N + a \pmod{N}$	Quick and infallible	none

Figure 2. *A comparison of the Agrawal-Kayal-Saxena primality test and the Fermat test. The AKS test is a fairly minor modification of the Fermat test but has one important advantage: it never gives wrong answers. By contrast, some numbers "fool" the Fermat test, as shown in Figure 1. The AKS test is the first known infallible test to run in polynomial time.*

In 1976, Gary Miller, who was at that time a graduate student at the University of California at Berkeley, showed that a variant of Fermat's little theorem *is* an all-purpose, polynomial-time, prime-proving test. Miller's result, however, relies on an unproved assumption known as the Generalized Riemann Hypothesis. The Riemann Hypothesis is a famous open problem in number theory regarding properties of a complex function known as the Riemann zeta function (see "A Prime Case of Chaos," *What's Happening in the Mathematical Sciences*, Volume 4, and "Think and Grow Rich," *What's Happening in the Mathematical Sciences*, volume 5). Basically it asserts that all the (nontrivial) zeroes of the zeta function lie along a particular line in the complex plane. The Generalized Riemann Hypothesis is a similar statement for an entire class of zeta functions, of which the Riemann zeta function is the simplest. Number theorists believe these hypotheses are true, but have yet to see a proof that withstands close scrutiny.

In 1980, Michael Rabin at Hebrew University of Jerusalem showed that Miller's test can be used as a polynomial-time *random* algorithm for identifying composite numbers. Miller's test basically does a collection of Fermat-style exponentiations mod N of various values a, and if the calculation "fails" for any a, then N is definitely composite. What makes the test polynomial-time is that the Generalized Riemann Hypothesis implies that every composite number fails for some relatively small value of a, so that a brief trial-and-error search will find it—and if it doesn't, then N must be prime. Rabin showed that if N is composite, then the calculation fails at least three-quarters of the time. Consequently, by simply picking values of a *at random* and running Miller's test, the probability of not detecting compositeness can be made arbitrarily small.

The Miller–Rabin test is extremely good at exposing composite numbers: The probability of a composite number passing the test for, say, 50 randomly chosen values of a is less than $1/2^{100}$, an extremely small number. On the other hand, passing the test for even a thousand values of a does not *prove* N to be prime, it merely makes it highly likely. Finding an efficient proof of primality was first addressed in 1986, by Shafi Goldwasser and Joe Kilian at M.I.T. They found a calculation based on randomly chosen elliptic curves which, with high probability, produces a proof of primality. At around the same time, Arthur Atkin at the University of Illinois at Chicago devised a different primality-proving algorithm also based on the theory of elliptic curves.

Atkin's algorithm was implemented by Francois Morain at the Institut National de Recherche en Informatique et en Automatique (INRIA) in France. Morain's program, ECPP (Elliptic Curve Primality Proving) is a mainstay of primality proving. In 2004, for example, it proved the primality of $2638^{4405} + 4405^{2638}$, a number with 15071 digits. ECPP and the Miller–Rabin test are an effective pair: If you're wondering if some large number is prime or composite, first test it with Miller–Rabin, which will almost certainly tell you if it's composite, and then (if Miller–Rabin says it's probably prime) run ECPP, which will almost certainly produce a proof of its primality.

Still, there is the element of uncertainty in Miller–Rabin and ECPP. Not that either test will ever give a wrong answer, but rather that they might offer *no* answer—there is the theoretical possibility that a run of Miller–Rabin will say your number N is probably prime but ECPP will fail to find a proof for it (so it will claim N is probably composite). In practice this is not a problem: If the tests don't provide a definitive answer, simply run them again—the odds against not getting an answer twice in a row are astronomically small. Nevertheless, number theorists and computer scientists continued to look for a non-random primality test that would run in polynomial time.

That's what the algorithm of Agrawal, Kayal, and Saxena succeeds in doing. Their test is yet another elaboration on Fermat's little theorem. Instead of exponentiating numbers (as in $a^N \equiv a$ mod N), it exponentiates polynomials. The key formula it uses is $(x + a)^N \cong x^N + a$ mod N for prime N. The familiar binomial theorem says that $(x+a)^N = x^N + \binom{N}{1}x^{N-1}a + \binom{N}{2}x^{N-2}a^2 + \cdots + \binom{N}{N-1}xa^{N-1} + a^N$. The only way for this to reduce to $x^N + a$ for all a is for N to divide each of its binomial coefficients $\binom{N}{k}$ for $k = 1$ to $N-1$. But a theorem dating back to Leibniz in the seventeenth century says that this happens if *and only if* N is prime. (This is the second time in this volume of *What's Happening in the Mathematical Sciences* that Leibniz's theorem rears its beautiful head—see the Venn Diagram article, p. 40.)

Simply computing $(x + a)^N$ mod N, however, is impractical, even with the trick of repeating squarings, because the computation would have to keep track of all N coefficients in the expansion. The extra wrinkle that makes the computation tractable is the introduction of a cleverly chosen, small prime number r. The computation of $(x + a)^N$ is done mod N in the coefficients and mod $x^r - 1$ in the powers of x. That is, every power x^k with k greater than or equal to r (including x^N) is replaced by x^{k-r}, over and over again. It is a particularly delightful form of polynomial multiplication in which there are never more than r coefficients, no matter how many times you multiply. Agrawal, Kayal, and Saxena's main theorem established criteria for choosing r and a set of test values a under which the computation of $(x + a)^N$ reduces to $x^N + a$ only when N is prime. (Actually, their theorem allows N to be a power of a prime. But it is extremely easy, in terms of computational complexity, to check whether a given number is a perfect square, cube, etc. So this last caveat has no significant effect on the run time of their primality-testing algorithm.)

The AKS algorithm (named for the initials of its discoverers) was immediately hailed as a breakthrough. Its simplicity made it easy to verify that the algorithm works as advertised. (Indeed, other number theorists chided themselves for not having discovered the algorithm earlier. It is as if they had been so hypnotized by Fermat's little theorem that they had failed to consider the possibility of using Leibniz's theorem for a primality test.) The only place where advanced, "modern" number theory is required is in making the estimate of its run time effective—in other words, figuring out exactly how rapidly it increases with the number of digits.

Because the number of digits of a number in binary notation is about $\log_2(n)$, and because the algorithm runs in polynomial time, its run time should be bounded by $\log_2(n)^d$, for some exponent d. The best exponent that Agrawal, Kayal and Saxena could get using elementary methods was $d = 10.5$. Recently, Hendrik Lenstra and Carl Pomerance have announced a variation of the AKS test that uses more complicated polynomial congruences, but has a better bound on the run time, with $d = 6$, which seems at present to be the best possible with the AKS approach. (Remember that a lower exponent is better.) In other words, the time that Lenstra and Pomerance's algorithm takes to determine primality is roughly the sixth power of the number of digits.

Ironically, the AKS algorithm (even as refined by Lenstra and Pomerance) is not yet the most efficient primality test for practical use, and it may never be. The ECPP algorithm, mentioned earlier, has a run time bounded by the fourth power of the number of digits, and it "almost always" works. It may even be true that ECPP always works, but number theorists just don't know. It is the ironclad certainty of the AKS algorithm that makes it so attractive. To mangle an old aphorism, mathematicians would rather have a bird in hand than two birds in hand that are not *known* to be birds.

Minding the Gaps

While the AKS breakthrough was quickly verified, the other two major breakthroughs in number theory were more difficult to check—and one of them proved to contain a significant error. That result concerned the varying size of gaps between consecutive primes.

If prime numbers occurred in some regular fashion, number theory would be a much simpler subject, and parts of it would be trivial. Instead, they are sprinkled seemingly randomly among the integers, sometimes closer together and sometimes farther apart. Except for 2 and 3, consecutive primes must differ by a gap of at least 2 (because all the prime numbers after 2 are odd). Many examples of primes differing by 2, or "twin primes," can be found: 3 and 5, 5 and 7, 17 and 19, 10037 and 10039. Consecutive primes with large gaps are somewhat harder to find; but, for example, there is a gap of 500 between the consecutive primes 303,371,455,241 and 303,371,455,741. Every integer between these two numbers is composite.

Although specific examples are hard to come by, it is easy to show that gaps between consecutive primes can be arbitrarily large. Because every number up through N is a divisor of $N! = N \times N - 1 \times \cdots \times 3 \times 2 \times 1$, the following $N - 1$ numbers are all composite: $N! + 2, N! + 3, \ldots, N! + N$. (To see this, note that 2 is a divisor of $N! + 2$, 3 is a divisor of $N! + 3$, and so on.) Hence the two primes separated by this sequence (the last prime less than $N! + 2$ and the first prime greater than $N! + N$) must be separated by a gap of at least N.

One of the most famous theorems of number theory, aptly called the Prime Number Theorem, provides excellent information on the *average* size of prime gaps. Proved independently by Jacques Hadamard and Charles de la Vallee Poussin in 1896, this theorem says that the number of primes less than or equal to N is approximately $N/\ln N$, where $\ln N$ is the natural logarithm of N. One implication is that the average gap between consecutive primes of size N is roughly $\ln N$. Alternatively, if each prime is "normalized" by dividing it by its logarithm, then the average gap between "normalized primes" is 1 (see Box, "The Gap Gap," p. 60). This theorem is one of the true triumphs of analytic number theory; it shows that in spite of the simple definition of prime numbers in terms of elementary arithmetic, they have some properties that cannot be understood without recourse to the tools of calculus, such as the natural logarithm.

Ironically, the AKS algorithm (even as refined by Lenstra and Pomerance) is not yet the most efficient primality test for practical use, and it may never be.

Unfortunately, our control over the average behavior of prime gaps tells us nothing about the frequency of large gaps or small gaps. Primes are too capricious for that. On the small-gap side, one possibility is that the smallest possible gap of 2 keeps on occurring over and over; in other words, there are infinitely many "twin primes." This is known as the Twin Primes conjecture. A more general conjecture, due to the French mathematician Alphonse de Polignac in 1849, asserts that, for *every* even number, there are infinitely many examples of consecutive primes that differ by that amount. Polignac's conjecture has not been proved for any even number, even if the primes are not required to be consecutive. Indeed, for all anyone knows, the opposite may be true. The difference between consecutive primes (and thus between any two primes) may grow larger and larger as the primes themselves increase.

One way to begin attacking these problems is to rephrase them somewhat. The Prime Number Theorem suggests that it would be useful to "normalize" each prime number by dividing it by its natural logarithm. This shrinks the average gap between "normalized primes" down to 1. Specifically, number theorists define the kth normalized gap to be $\Delta_k = (p_{k+1} - p_k)/\ln p_k$, where p_k is the kth prime (see Box, "The Gap Gap"). For instance, 303,371,455,241 is the 11,945,986,787-th prime and 303,371,455,241 is the 11,945,986,788-th prime. Thus, $\Delta_{11,945,986,787)} \approx 18.2$. This confirms our suspicion that the gap between these two primes is unusually large. However, in 1931, an otherwise obscure mathematician named Erik Westzynthius showed that there is no upper bound on the size of the normalized gaps Δ_N.

How about the other end of the spectrum: How small can the gaps Δ_N get? If the Twin Prime conjecture—or virtually any variant of it—is correct, then there are gaps that are arbitrarily small. That would be the "easiest" way to settle the question. For the time being, number theorists consider the Twin Prime conjecture to be far beyond their reach. Consequently, they have focused on the smallness of normalized gaps, hoping in this way to gradually build up their knowledge for an assault on the Twin Prime conjecture.

For the time being, number theorists consider the Twin Prime conjecture to be far beyond their reach. Consequently, they have focused on the smallness of normalized gaps, hoping in this way to gradually build up their knowledge for an assault on the Twin Prime conjecture.

The official object of study is the "limit infimum" of the sequence of gaps, written $\underline{\Delta} = \liminf \Delta_N$. This requires a modicum of explanation. What number theorists are interested in is not the absolute minimum value ever achieved by the sequence of Δ_N's (which is the negative value, Δ_2), but rather the smallest value that is approached by infinitely many terms of the sequence. In other words, it doesn't matter what happens for "small" values of N, what's important is the asymptotic behavior. The limit infimum of a sequence is the smallest number that is the limit of some subsequence (see Figure 3).

Consider the sequence:
3, 2, 3.1, 1.5, 3.14, 1.25, 3.141, 1.125, 3.1415, 1.0625, 3.14159,...
The odd-numbered terms (in red) converge to Pi and the even-numbered term (in blue) converge to 1, hence the limit infimum of the sequence is 1

Figure 3. *The limit infimum of a sequence is the smallest limit of any subsequence.*

The only "obvious" fact about $\underline{\Delta}$ is that it cannot be larger than 1. But that's only obvious because of the Prime Number Theorem. (One might think that Westzynthius's theorem, which shows there are infinitely many gaps larger than 1, would imply there are also infinitely many smaller than 1, in order to get an overall average of 1, but that's not correct. It's possible for the average value of an infinite sequence to be 1 even if all of its terms are greater than 1, as the example 1.1, 1.01, 1.001, 1.0001, etc., shows.)

The first step toward showing that $\underline{\Delta} = 0$ was taken by the Hungarian mathematician Paul Erdős (see "Proof by Example: A Mathematician's Mathematician," *What's Happening in the Mathematical Sciences*, Volume 4) in 1940, who showed that $\underline{\Delta} < 1$. Twenty-six years later, Harold Davenport at the University of Cambridge and Enrico Bombieri, then at the University of Pisa, dramatically improved Erdös's result, to $\underline{\Delta} < (2+\sqrt{3})/8 \approx 0.4665$. (In 1926, the British mathematicians G.H. Hardy and John Littlewood proved that $\underline{\Delta} \leq 2/3$, but, like Gary Miller's test for primality, their proof assumed the Generalized Riemann Hypothesis.) This was improved by several number theorists over the next two decades, reaching $\underline{\Delta} < 0.2486$, proved by Helmut Maier in 1984.

Then came Goldston and Yildirim's announcement. At a conference in March 2003, held in Olberwohlfach, Germany, the

Even if Pintz had not found a way to make things work, Goldston and Yildirim's original work would have still been worthwhile. Just ask Ben Green and Terence Tao.

two number theorists reported on a proof that $\Delta = 0$. In fact, they had a stronger version: The difference between consecutive primes, $p_{N+1} - p_N$, is infinitely often less than $(\ln p_N)^{8/9}$. This is a far cry from what the Twin Prime conjecture implies, which is that the difference is infinitely often equal to 2, but it is so much stronger than the previous results that it came as a bit of a shock to other number theorists.

Other researchers immediately began poring over Goldston and Yildirim's proof. Mathematicians often find minor mistakes in each other's work that are easy to correct; equally often they find ways to simplify proofs or generalize the results. In this case, however, an error became apparent that was not so easy to fix. Andrew Granville at the University of Montreal and Kannan Soundararajan at the University of Michigan spotted a problem in one of the crucial estimates in Goldston and Yildirim's proof. They became suspicious, Granville recalls, when they realized that the estimate actually implied far more than Goldston and Yildirim had claimed: If true, it meant that the difference between consecutive primes is infinitely often less than or equal to 12. This struck them as altogether too good to be true, so they looked more closely at the derivation of the estimate—and discovered that it wasn't correct.

The error lay in a technical lemma. Without it, Goldston and Yildirim were still able to get an improvement over previous results on Δ. But they were no longer able to show it was equal to 0. A year later, Goldston had a new idea for achieving the breakthrough, but it also fell short. Then, in December 2004, he was contacted by Janos Pintz, a number theorist at the Renyi Mathematical Institute in Hungary, who saw a way to fix the problem in Goldston's new approach. By February 2005, the three mathematicians had a new—and considerably simpler—proof.

Given the fate of the first proof, Goldston, Yildirim, and Pintz decided to circulate the new proof to a small group of experts, including Granville and Soundararajan, to see if it would withstand scrutiny. Not only did it hold up, but one of the experts, Yoichi Motohashi at Nihon University in Japan, found additional ways to simplify the proof that $\Delta = 0$.

Even if Pintz had not found a way to make things work, Goldston and Yildirim's original work would have still been worthwhile. Just ask Ben Green and Terence Tao.

Prime Gaps and Killer APPs

As noted before, prime numbers are capricious. One way to test their capriciousness is to look for unusually long or short gaps between consecutive primes. But there is a second way, which involves looking for evenly-spaced sequences of primes that are not necessarily consecutive. These can be thought of as short bursts of order within the randomness. Green and Tao electrified the mathematical world in April 2004, with a stunning announcement: The sequence of prime numbers in fact contains arbitrarily long "bursts of order." Part of their proof used some of the (unflawed) estimates Goldston and Yildirim had derived in their work on prime gaps.

An arithmetic progression (the accent in "arithmetic" is on the syllable "met") is a sequence (either finite or infinite) of integers that differ by a constant amount, such as 7, 37, 67, 97, 127, 157, 187, etc., which differ by 30. The first six terms in this example are all primes, making it what's called an arithmetic progression of primes (or APP, for short) of length 6. Other examples of APPs include {11, 41, 71, 101, 131} (length 5), {13, 43, 73, 103} (length 4), and so on.

An arithmetic progression of primes cannot be infinitely long, for two reasons. First, if p is the kth prime in a progression with difference d, then the $(p + k)$th term is $p + dp$, which is divisible by p. Second, if q is the smallest prime not dividing d, then q necessarily divides one among the first q terms in the progression (and every qth term thereafter). Thus an APP cannot be longer than the smallest prime it contains, nor longer than the smallest prime not dividing its constant difference. This is why the examples given above go up in increments of $30 = 2 \times 3 \times 5$. If we wanted to generate longer APPs, we would need increments of at least $210 = 2 \times 3 \times 5 \times 7$, and we would need to start with primes larger than 7.

Even though there are no infinitely long arithmetic progressions of primes, number theorists have long suspected there are APPs of any given finite length. The evidence was admittedly thin: The longest APP yet found has only 23 terms. Its initial prime is 56,211,383,760,397, and its constant difference is 44,546,738,095,860 ($2^2 \times 3 \times 5 \times 7 \times 11 \times 13 \times 17 \times 19 \times 23 \times 99839$). It was found in 2004 by computer programmer Markus Frind and fellow prime number enthusiasts Paul Underwood and Paul Jobling. (Frind and others had earlier found APPs with 22 terms.) Nevertheless, number theorists felt confident that longer and longer progressions continue to exist. They just weren't sure they'd ever be able to prove it.

It's hard to prove the existence of arithmetic progressions anytime the set involved lacks an obvious additive structure, as is the case for the prime numbers. Nevertheless, there are some notable examples. In 1927, the Dutch mathematician Bartel van der Waerden gave the first such proof. Imagine the (positive) integers are partitioned into a finite number of sets—you can think of each number as being assigned one of several colors. Van der Waerden showed that no matter how the numbers are partitioned, at least one of the sets contains arbitrarily long arithmetic progressions. Because the assignment of colors can be completely arbitrary, van der Waerden's theorem is saying, in a certain sense, that even random sets of integers have some bursts of order in them if they are dense enough.

In 1975, the Hungarian mathematician Endre Szemeredi proved a much stronger version of van der Waerden's theorem. Szemeredi showed that the only way an infinite set of integers can avoid having arithmetic progressions of arbitrary length is if its "asymptotic density" as a subset of the integers is 0. The density of any finite string of positive integers $b_1 < b_2 < \ldots < b_n$ is simply n/b_n. The asymptotic density of an infinite set $b_1 < b_2 < \ldots < b_n < \ldots$ is simply the limit of n/b_n as n tends to infinity. This limit may or may not exist. (If it doesn't, one can work with the "limit supremum," which is similar to the limit infimum described earlier.) Szemeredi's theorem says that in order for a bunch of b_n's to avoid containing arithmetic progressions of arbitrary length, the limit must exist and must equal 0. In other words, any subset of integers of *positive* density necessarily has arithmetic progressions of arbitrary length.

Ben Green. *(Photo courtesy of Ben Green.)*

Unfortunately, the primes do not have positive density—the Prime Number Theorem implies that $p_n/n \approx 1/\ln n$, which tends to 0—so Szemeredi's theorem does not apply. The best that was known, a result first proved by van der Corput in 1939, is that there are infinitely many examples of APPs of length 3. (Anytime there are arithmetic progressions of arbitrary length there are automatically infinitely many progressions of any given length.) The next best result, proved by Roger Heath-Brown in 1981, is the same statement for progressions of length 4, except that the fourth number is allowed to be either a prime number or the product of just two primes.

Green and Tao thought they might be able to force the fourth number in Heath-Brown's theorem to be prime instead of "almost prime." In 2001, Green's Ph.D. advisor, Timothy Gowers, had found a new, more powerful proof of Szemeredi's theorem based on techniques from harmonic analysis. Green and Tao thought these techniques might be enough to solve the 4-term problem. Instead, they got the whole nine yards.

Green and Tao's proof is a tour de force of analytic number theory, combining techniques from harmonic analysis and ergodic theory. In a broad outline, the proof has three key components. First, Green and Tao proved a general "transference principle" for Szemeredi's theorem, which allows it to apply in a so-called "pseudo-random" setting. They next proved that the set of "almost primes"—meaning numbers that have very few prime factors relative to their size—is such a setting. Finally—and this was the step that utilized portions of Goldston and Yildirim's work—they showed that the primes have positive density within the set of almost primes.

The theorem is surprisingly strong. It implies, for example, that one can ignore, say 99% of the primes and still guarantee arithmetic progressions of arbitrary length. Even so, number theorists still have grander goals to shoot for. Erdős, who was known for offering cash prizes for solutions to mathematical problems, offered \$3000 for a proof that arithmetic progressions of arbitrary length exist in any sequence of positive integers $b_1 < b_2 < \ldots < b_n < \ldots$ for which the sum of reciprocals, $1/b_1 + 1/b_2 + 1/b_3 + \cdots$, diverges. This would entail the result for APPs, in light of Euler's famous proof for the reciprocals of the primes.

Terence Tao. *(Photo courtesy of Terence Tao.)*

It is fair to say, though, that no mathematician studies prime numbers for the money. The chance to unveil new secrets of one of the oldest and most mysterious concepts in math is reward enough.

Carl Friedrich Gauss. *Carl Friedrich Gauss as a young man. (Photo courtesy of Northwestern State University of Louisiana, Watson Memorial Library, Cammie G. Henry Research Center.)*

From Rubik's Cube to Quadratic Number Fields... and Beyond

Dana Mackenzie

I T ALL STARTED WITH A RUBIK'S CUBE. Not the original $3 \times 3 \times 3$ version, but a simpler $2 \times 2 \times 2$ "mini-cube" that Manjul Bhargava had in his dormitory room. The Princeton graduate student was spending the summer of 1998 in California, and thinking about how to count algebraic number fields. It wasn't a new problem. The concept of a number field (to be explained below) dates back, in some sense, to the number theorist's bible, the classic 1801 treatise by Carl Friedrich Gauss called *Disquisitiones Arithmeticae* (see Figure 1, next page).

Gauss had discovered a strange way of combining two quadratic polynomials, of a type known as "quadratic forms," to get a third quadratic form. Number theorists call the rule "Gauss composition," and they have used it for two centuries without really improving upon its original definition. But as he thought about his Rubik's mini-cube one June day, Bhargava suddenly realized that he was staring at a better version of Gauss composition. "I was visualizing numbers on each of the corners," Bhargava said. "I saw these binary quadratic forms coming out, three of them. I just sat right down and wrote out the relations between them. It was a great day."

What Bhargava had done was break number theorists out of a two-century-old way of thinking about quadratic forms in terms of matrices, which are flat, two-dimensional arrays of numbers. Every math major learns about matrices in college (and many non-math majors do as well). They are one of the most indispensable tools of mathematics. By contrast, you will never hear about three-dimensional "boxes" of numbers, even one as simple as a $2 \times 2 \times 2$ box, in an undergraduate math course. You would have to look very hard to find them mentioned in a graduate course. Even most mathematicians wouldn't know what to do with them.

But what Bhargava discovered that day, and continued to discover during a feverish summer of work, was that boxes have a simple theory that is every bit as elegant as the number theory of matrices. Gauss composition, which looks mysterious when phrased in terms of matrices, can be expressed beautifully in terms of boxes. Not only that, Bhargava discovered other forms of composition that act on various other types of polynomial forms—laws that had never even been suspected before. Bhargava had stumbled on the keys to a new kingdom.

Manjul Bhargava. *(Photo courtesy of Denise Applewhite, Princeton University, Office of Communications.)*

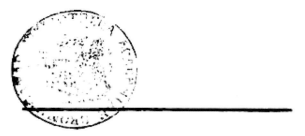

DISQVISITIONES

ARITHMETICAE

AVCTORE

D. CAROLO FRIDERICO GAVSS

LIPSIAE

IN COMMISSIS APVD GERH. FLEISCHER, JVN.

1 8 0 1.

Figure 1. *Gauss's* Disquisitiones Arithmeticae *essentially founded the discipline of number theory. In this book Gauss defined a "composition law" of quadratic forms that Bhargava has now generalized to cubic and higher-degree forms.*

Bhargava's doctoral dissertation, finished in 2001—the 200th anniversary of Gauss's *Disquisitiones*—has already received high accolades. He was the very first recipient of a new five-year fellowship from the Clay Mathematics Institute; he received the 2005 Blumenthal Award for research in pure mathematics from the American Mathematical Society (AMS); and in 2003, just two years after receiving his doctorate, he became one of the youngest full professors in Princeton's history, at age 28. "Manjul's thesis was quite magical," says his advisor, Andrew Wiles. "With completely elementary tools, he created a whole new theory that extended the work of Gauss in an totally unexpected way."

For anyone else, such a meteoric career path might seem like a surprise. For Bhargava, it seems effortless. He has a unique ability to think about age-old problems in a new way. As a college student, he came up with a way to generalize the classical factorial function, and for this work he won the AMS's Morgan Prize for undergraduate research. When he was in graduate school, one of his professors, John Horton Conway, mentioned to him an unpublished theorem on quadratic forms, called the Fifteen Theorem, that he had proved several years earlier with one of his students, William Schneeberger. Within a few months, Bhargava had a better proof, which (as Conway wrote) "has made it unnecessary for us to publish our rather more complicated proof."

Although not related to his work on Gauss composition, the Fifteen Theorem makes a good starting point to introduce some of the concepts involved. Ever since 1640, when Pierre de Fermat proved a simple description of all the whole numbers that are expressible as a sum of two squares (such as $5 = 1^2 + 2^2$), number theorists have been fascinated by "representations" of numbers. Which numbers, for instance, can be written as a square plus twice a square, $x^2 + 2y^2$? Or as a sum of two cubes, $x^3 + y^3$? (This was the line of thinking that led Fermat to his celebrated "Last Theorem.") Or as a sum of three squares, $x^2 + y^2 + z^2$?

In 1770, Joseph Louis Lagrange proved that four squares suffice for representing every positive integer. Much later, in 1910, Srinivasa Ramanujan provided a list of 54 additional quadratic polynomials in four variables that represent every integer, such as $x^2 + y^2 + z^2 + 2w^2$ and $x^2 + 2y^2 + 4z^2 + 7w^2$. His list had one inadvertent mistake. Ramanujan included $x^2 + 2y^2 + 5z^2 + 5w^2$ on the list, but actually the number 15 cannot be expressed in this form. (To see this, note that z and w would both have to be 0 or 1. If they are both 0, then $x^2 + 2y^2 = 15$, but there are no integers x and y that solve this equation. If one of z or w is 1, $x^2 + 2y^2 = 10$, which is again unsolvable in integers. If both z and w are 1, then $x^2 + 2y^2 = 5$, which is also unsolvable.)

Ramanujan's list looked somewhat haphazard, but Conway had realized that there is a system to it. The polynomials in Ramanujan's list are all called quadratic forms: they are quadratic polynomials in four variables, in which *every term* is quadratic. Quadratic forms can always be written in terms of symmetric matrices, using a matrix product. For example,

$$x^2 + 2y^2 + 5z^2 + 5w^2 = (x, y, z, w) \begin{bmatrix} 1 & 0 & 0 & 0 \\ 0 & 2 & 0 & 0 \\ 0 & 0 & 5 & 0 \\ 0 & 0 & 0 & 5 \end{bmatrix} \begin{pmatrix} x \\ y \\ z \\ w \end{pmatrix}.$$

If all the entries in the matrix are integers, then the form is called an integer matrix form. Another way of saying this is that the coefficients of any cross terms such as xy, xz, etc., must all be even. (For the forms in Ramanujan's list, the cross terms are all zero.) Conway and Schneeberg's Fifteen Theorem makes it very easy to tell whether an integer matrix form in four variables represents every integer. You only need to check the integers from 1 to 15. Thus, to check that every integer is a sum of four squares, you only have to do it for the first fifteen: $1 = 1^2 + 0^2 + 0^2 + 0^2; 2 = 1^2 + 1^2 + 0^2 + 0^2; 3 = 1^2 + 1^2 + 1^2 + 0^2;$ and so on through $15 = 1^2 + 1^2 + 2^2 + 3^2$. If an integer matrix form fails to represent all the integers, as $x^2 + 2y^2 + 5z^2 + 5w^2$ did, it will fail early.

What if the form is allowed to have odd coefficients for terms like xy, xz, etc.? This type of form is called an integer-valued form. Numerical experiments convinced Conway that, in the more general case of integer-valued forms, the Fifteen Theorem becomes the "290 theorem." If the form succeeds in representing every integer up through 290, then it will represent every integer. But Conway was unable to prove it, and he dangled the problem in front of Bhargava as a challenge.

Bhargava took the bait. "It was a nice distraction from my thesis writing," he says. "It got me so interested that I abandoned my thesis for a few months, much to the dismay of my advisor." But his "distraction" paid off. Not only did Bhargava prove the Fifteen Theorem, but he found a host of similarly intriguing and unexpected results. For instance, an integer matrix form represents all odd numbers if and only if it represents 1, 3, 5, 7, 11, 15, and 33. Finally, after several years of work, he and Jonathan Hanke polished off the proof of the 290 Theorem in 2005, and in this way completed the challenge set out by Ramanujan and generalized by Conway. It turns out that there are 6436 integer-valued quadratic forms in four variables that represent all the integers. Of these, Ramanujan had identified all the ones which have no cross terms.

Fortunately, Bhargava's "distraction" by this set of problems didn't keep him from finishing his thesis—and more.

The goal of Bhargava's thesis, and much of his postdoctoral research as well, is to develop ways for mathematicians to count objects of number-theoretic interest. Although it sounds simple, counting can be hard work. (Just ask the Census Bureau.) Many an important mathematical theorem has boiled down to finding a way to count a set, or sometimes finding two different ways to count the same set. If a set has N elements, and you have two different expressions that count it, then the two expressions have to be equal. A good example of this is the very first mathematical problem that Gauss ever solved. According to legend, his arithmetic teacher told his class to add the numbers from 1 to 100, expecting the exercise to keep the class quiet a long time. Gauss almost immediately walked to the front of the classroom with the answer written on his slate. He had found a different way to add the numbers up. He added them in pairs: $(1 + 100), (2 + 99), (3 + 98)$, and so on. There are 50 pairs, each adding to 101, so the sum is 5050.

There are many counting arguments just like this in number theory. A modern mathematician would recognize Gauss's argument as a way of counting the number of lattice points inside a triangle (see Figure 2). If you are given a lattice in the plane—which can be thought of as the set of ordered pairs (x, y) such that both x and y are integers—and if you describe a region in the plane with a finite number of polynomial inequalities (such as $x > 0, x \le y, y \le 100$), it is relatively simple to count the number of lattice points inside that region. By analogy, the hard way to take a census is to go to everybody's house and count them. The easy way is to get everyone to come to a large open space, stand in an array and work out the total population from the dimensions of the array. (The new "hard part" is making sure that everyone actually came.)

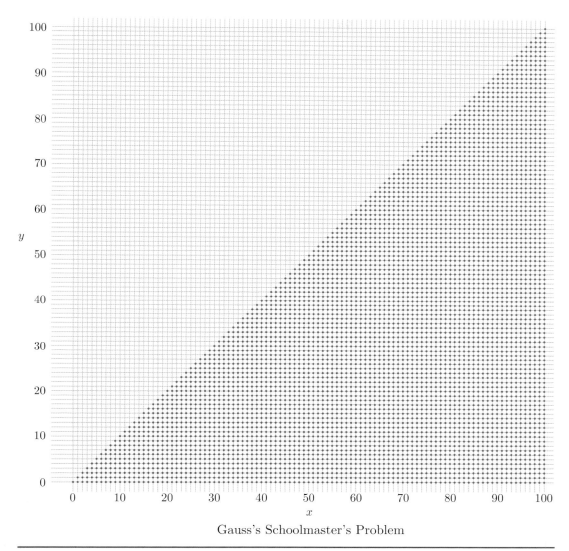

Gauss's Schoolmaster's Problem

Figure 2. *The problem posed by Gauss's schoolmaster—finding the sum of the numbers from 1 to 100—can be rephrased as finding the number of lattice points in a region in the plane whose boundary is defined by polynomial equations. (Here the boundaries are simply line segments.) Bhargava's work translates complicated number theory problems into the same type of lattice point problem, possibly in several dimensions.*

This counting technique holds true for higher-dimensional lattices as well. For example, estimating the number of solutions to $x^2 + y^2 + z^2 + w^2 \leq 290$ is easy. It amounts to finding the number of lattice points (x, y, z, w) inside a four-dimensional sphere. Counting the exact number of solutions to $x^2 + y^2 + z^2 + w^2 = 290$ (that is, finding the number of lattice points *on* the sphere) is somewhat harder. In general, counting lattice points that lie on a particular curve or surface is a trickier problem than counting lattice points inside a region.

The central project behind Bhargava's thesis, which has expanded to four lengthy papers in *Annals of Mathematics* and counting, is to count algebraic number fields and various interesting structures on them. The approach in every case is the same: find a way to make the fields and structures correspond to lattice points in an n-dimensional vector space. Bhargava has found more than twenty different examples, only four of which were previously known.

An *algebraic number field*, or number field for short, is obtained by adding or "adjoining" a finite number of solutions to polynomial equations to the standard field of rational numbers. Within any number field there is a "ring of integers" which have properties very much like ordinary integers. Because the subject of number theory dealt originally with the properties of ordinary integers, it has naturally expanded its scope over the years to include the properties of algebraic integers and number fields as well.

The prototypical example of a number field is obtained by adjoining the imaginary unit $i = \sqrt{-1}$ (a solution to the polynomial equation $x^2 + 1 = 0$) to the rational numbers. By the rules of addition and multiplication, this means that all the numbers of the form $a + bi$, where a and b are rational numbers, must be included in the number field as well. It turns out that no more are needed: this collection forms a field. The integers in this field (known as the Gaussian integers) are just the ones you would expect: they are numbers of the form $a + bi$, where a and b are ordinary integers.

Even if you are a pure number theorist who only cares about ordinary integers, you will soon get sucked into studying algebraic integers because they are so convenient. For example, Fermat, when proving his theorem about numbers that were representable as sums of squares $(x^2 + y^2)$, did not use Gaussian integers because the concept of $\sqrt{-1}$ was not acceptable at the time. As a result, his proofs were very laborious. The same results could have been proven much more simply if he had been able to factor $x^2 + y^2$ as $(x + iy)(x - iy)$ and work in the ring of Gaussian integers. Even Gauss, working in 1801, was reluctant to invoke the still-novel complex numbers. "At that time the fashion was to say anything about the integers purely in terms of the integers," says Bhargava. "People would do secret calculations on scratch paper outside the integers and then translate it all back into the integers. Gauss was likely one of those people. Some of his calculations are extremely long and laborious, and they would have been much simplified if he had written them in the language of quadratic fields."

However, algebraic rings do need to be handled with care. In the Gaussian integers, we can define prime numbers, factor integers into primes, and prove that the decomposition is unique, just as in the ordinary integers. But not all number fields are so well behaved. For instance, if we had chosen to add $\sqrt{-5}$ to the rational numbers, we would get algebraic integers of the form $a + b\sqrt{-5}$ (where a and b, again, are ordinary integers). This ring of integers has no catchy name; it is simply denoted $Z[\sqrt{-5}]$ (read, "Z-adjoin-square-root-of-minus-five"). In this ring, the number 6 can be factored in two different ways: $6 =$

$2 \cdot 3$, and $6 = (1 + \sqrt{-5})(1 - \sqrt{-5})$. All of these factors are prime in this ring, and hence the uniqueness of prime factorization—one of the most natural, seemingly obvious properties of the natural numbers—fails in $Z[\sqrt{-5}]$.

Number theorists were perplexed by this phenomenon at first. But Richard Dedekind and others salvaged something out of the situation by defining what they called "ideal numbers" (later shortened to "ideals"), for which a version of unique factorization does hold. "Ideal numbers" can be thought of as a special kind of sub-lattice of the lattice of integers. One example in $Z[\sqrt{-5}]$ is the ideal $\langle 1, \sqrt{-5} \rangle$, consisting of all numbers that can be written as $a + b\sqrt{-5}$. (This ideal is, obviously, the whole ring.) Another example is the ideal $\langle 2, 1 + \sqrt{-5} \rangle$, which consists of all numbers that can be written as $2a + (1 + \sqrt{-5})b$ for some integers a, b. The two ideals give rise to different lattices. Even more important, from the point of view of ring theory, is that every other ideal in $Z[\sqrt{-5}]$ is just a re-scaled version of one of these two. In other words, you can get every other ideal by multiplying all the elements of one of these lattices by some fixed number $x + y\sqrt{-5}$. Thus the ideals split into two ideal classes, like a group of children splitting into two teams. There are the "friends of $\langle 1, \sqrt{-5} \rangle$" (called the principal ideal class) and the "friends of $\langle 2, 1 + \sqrt{-5} \rangle$" (see Figure 3).

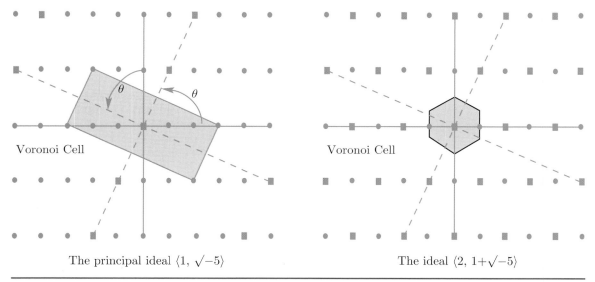

Voronoi Cell

The principal ideal $\langle 1, \sqrt{-5} \rangle$

Voronoi Cell

The ideal $\langle 2, 1+\sqrt{-5} \rangle$

Figure 3. *Ideals in $Z[\sqrt{-5}]$ are lattices in the complex plane with an additional property of closure under multiplication by $\sqrt{-5}$ (in other words, rotation by 90° followed by a magnification by $\sqrt{5}$). Principal ideals (left) are just rescaled versions of the integer lattice in $Z[\sqrt{-5}]$. Non-principal ideals (right) have a different shape, as one can see by drawing the Voronoi cell (the set of points that lie closer to the origin than to any other point in the ideal). Thus these two ideals belong to different ideal classes. The algebra of ideal classes underlies Gauss's law of composition and Bhargava's generalization.*

The same setup can be repeated in any number field. Every number field has a ring of integers, ideals, a principal ideal (the ring itself), and ideal classes. The rings with only one ideal class, in which every ideal is a re-scaled version of the principal ideal, have unique factorization. The Gaussian integers are one example. The rings with more than one ideal class, such as $Z[\sqrt{-5}]$, never have unique factorization. The number of ideal classes is, in some sense, a measure of how badly unique factorization fails, and so this number (called the "class number") is one of the most important pieces of information about a field. The class number of $Z[\sqrt{-5}]$ is two.

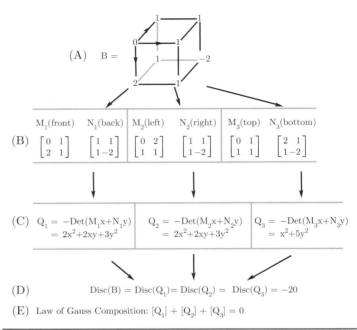

(A)   B =

(B) M_1(front) N_1(back) M_2(left) N_2(right) M_3(top) N_3(bottom)

$$\begin{bmatrix} 0 & 1 \\ 2 & 1 \end{bmatrix} \quad \begin{bmatrix} 1 & 1 \\ 1 & -2 \end{bmatrix} \quad \begin{bmatrix} 0 & 2 \\ 1 & 1 \end{bmatrix} \quad \begin{bmatrix} 1 & 1 \\ 1 & -2 \end{bmatrix} \quad \begin{bmatrix} 0 & 1 \\ 1 & 1 \end{bmatrix} \quad \begin{bmatrix} 2 & 1 \\ 1 & -2 \end{bmatrix}$$

(C) $Q_1 = -\text{Det}(M_1 x + N_1 y)$ $Q_2 = -\text{Det}(M_2 x + N_2 y)$ $Q_3 = -\text{Det}(M_3 x + N_3 y)$
$\quad\quad = 2x^2 + 2xy + 3y^2$ $= 2x^2 + 2xy + 3y^2$ $= x^2 + 5y^2$

(D) $\text{Disc}(B) = \text{Disc}(Q_1) = \text{Disc}(Q_2) = \text{Disc}(Q_3) = -20$

(E)   Law of Gauss Composition: $[Q_1] + [Q_2] + [Q_3] = 0$

Figure 4. *Bhargava's version of Gauss's law of composition, using 2 x 2 x 2 number boxes. (a) A number box consisting of integers. (b) Three different ways to slice the box into two number squares (or matrices): front and back, left and right, or top and bottom. [Note: You have to turn your head to read the left-hand matrix in such a way that 0 is in the top left corner.] (c) Each pair of matrices corresponds to a quadratic form as shown. (d) Each quadratic form has the same discriminant. (e) The Gauss composition law. Amazingly, Bhargava showed that the arrows can be reversed. That is, if three quadratic forms have the same discriminant and [Q1] + [Q2] = −[Q3], then all three quadratic forms come from one number box.*

Finding the class number is hard. Formulas do exist for it, but they look more like black magic than mathematics. Gauss developed a much more straightforward approach in his *Disquisitiones*, even though the language of ideals had not been developed yet. Instead of ideals, he used the language of quadratic forms. Any integer matrix form, $ax^2 + bxy + cy^2$ with b even, has a discriminant, given by $b^2 - 4ac$. This is the same discriminant learned about in high school algebra. Each ideal class of the ring $Z[\sqrt{d}]$ corresponds to a unique quadratic form with discriminant $4d$, and vice versa. (For quadratic forms with b odd, a very similar statement is true, but for simplicity let us skip this case.) Therefore the class number is the same as the number of essentially different quadratic forms. Because the Gaussian integers, $Z[\sqrt{-1}]$, have unique factorization, there is only one quadratic form of discriminant -4, namely $x^2 + y^2$. On the other hand, because $Z[\sqrt{-5}]$ does not have unique factorization, there are two quadratic forms of discriminant -20, namely $x^2 + 5y^2$ and $2x^2 + 2xy + 3y^2$. The whole point of the exercise is that quadratic forms are easier to count than ideal classes, because quadratic forms are simply described by the three integers (a, b, c). In other words, Gauss had figured out a way to take a census of ideal classes by lining them up in a lattice in three-dimensional space. Admittedly, the count is still not completely routine, because a specific discriminant has been specified (in this case, $b^2 - 4ac = -20$).

Actually, Gauss did something even more impressive: he showed that the quadratic forms have an operation on them, a sort of multiplication. Given any two quadratic forms with the same discriminant, you can compose them to find a third quadratic form that also has the same discriminant. For example, $[2x^2 + 2xy + 3y^2]$ composed with itself equals $[x^2 + 5y^2]$. The rule of composition looks quite bizarre to the uninitiated, but it works, and provides a nice way of making sure that you have gotten all the population out for the "census." Mathematically, it gives the set of "essentially different" quadratic forms (or, equivalently, the set of ideal classes) the algebraic structure of a group.

But why does each pair of quadratic forms lead to a third quadratic form? That was the question that Manjul Bhargava answered with his beautiful Rubik's cube construction. To see why, start with a $2 \times 2 \times 2$ cube of numbers, such as the one shown in Figure 4.

What Bhargava realized is that there are three ways to break down this cube into squares. You can split it into front and back:

$$M_1 = \begin{bmatrix} 0 & 1 \\ 2 & 1 \end{bmatrix}, N_1 = \begin{bmatrix} 1 & 1 \\ 1 & -2 \end{bmatrix}.$$

Or you can split it into left and right:

$$M_2 = \begin{bmatrix} 0 & 2 \\ 1 & 1 \end{bmatrix}, N_2 = \begin{bmatrix} 1 & 1 \\ 1 & -2 \end{bmatrix}.$$

Or you can split it into top and bottom:

$$M_3 = \begin{bmatrix} 0 & 1 \\ 1 & 1 \end{bmatrix}, N_3 = \begin{bmatrix} 2 & 1 \\ 1 & -2 \end{bmatrix}.$$

To get quadratic forms out of this, he combined the squares with variable coefficients x and y to form new matrices, $xM_i + yN_i$. Finally, he took the determinant of each 2×2 matrix, and in this way obtained three quadratic forms:

$$Q_1(x, y) = -\det(M_1 x + N_1 y)$$

$$= -\det \begin{bmatrix} y & x + y \\ 2x + y & x - 2y \end{bmatrix} = 2x^2 + 2xy + 3y^2$$

and, in similar fashion, $Q_2(x, y) = 2x^2 + 2xy + 3y^2$, $Q_3(x, y) = x^2 + 5y^2$. These are the same forms that appeared above in the law of Gauss composition! In fact, Bhargava showed this was no accident. Whenever you have three quadratic forms $[Q_1]$, $[Q_2]$, and $[Q_3]$ such that $[Q_1] + [Q_2] + [Q_3] = 0$ according to the law of Gauss composition, the three forms must have come from the same $2 \times 2 \times 2$ "box." The threefold nature of the group law comes directly from the three-dimensional nature of the box, which led to the three ways of decomposing it. By looking at only one quadratic form or one planar matrix at a time, mathematicians had missed the three-dimensional structure of the whole.

A pretty consequence of Bhargava's definition was that boxes themselves have a group law, and they also have a discriminant. The discriminant of the box is exactly the same as the discriminant of each of the three quadratic forms $Q_1, Q_2,$ and Q_3 (which are all equal). There is a definition of box equivalence that is very much like the notion of equivalent matrices under "row reduction," a standard topic of undergraduate courses in matrix algebra. Only now there are rows and columns and layers because of the three-dimensional nature of the box. The original box is considered equivalent to the new box with row 2 replaced by (row 2 + x row 1), where x is any real number. The same is true if column 2 is replaced by (column 2 + x column 1) or (layer 2 + x layer 1). This notion of equivalence preserves the discriminant (that is, any two equivalent boxes have the same discriminant). Furthermore, the discriminant is the *only* quantity that is left unchanged by all of these row, column, and layer operations. Bhargava's law of composition is the *only* operation on boxes that is compatible with Gauss composition. In that sense, there is nothing arbitrary about the way Bhargava defined discriminants and composition; it is the only way he could have done it.

One is left with the feeling that anyone in the two hundred years between Gauss and Bhargava who really took the trouble to understand Gauss composition, and who happened to think of boxes instead of squares, could have come up with the same idea. Bhargava doesn't dispute this. "I often say that I was born in the wrong century!" he says. "I would have fit well into the nineteenth century. All of these ideas are very simple, classical things that you could explain to a nineteenth-century mathematician."

In fact, the ideas were out there; they just had not been appropriated by number theorists. A "box" is more formally called a tensor product, and tensor products have been studied by geometers for years. Some of the other vector spaces that Bhargava uses to count number fields are "alternating" or skew-symmetric products, which have also been in use since the nineteenth century. A key invariant on these vector spaces, which Bhargava uses to produce higher composition laws, is called the Pfaffian. Another useful tool, used to identify so-called "simple" alternating products, is called the Plucker relation. One cannot help thinking that Julius Plucker (1801–1868) and Johann Pfaff (1765–1825, a friend of Gauss) would have been delighted to hear about this application of their work. "One of the most fun moments in my career was when the parametrization of quintic rings worked out," Bhargava says. "It is all about the Pfaffians of 10-by-10 matrices."

With his menagerie of lattices in exotic vector spaces from the nineteenth century, Bhargava has made it possible to attack problems that were previously untouchable. For example, perhaps the simplest counting question: given a number d, roughly how many number fields of discriminant d are there, on average? The answer for quadratic fields has been known for decades; on average, there are $6/\pi^2$ such number fields.

For cubic fields (number fields created by adjoining a solution of a cubic polynomial), the answer was obtained with great difficulty by Harold Davenport and Hans Heilbronn in 1971: the average is $1/\zeta(3)$, where ζ is the Riemann zeta-function. For quartic and quintic fields, Akihiko Yukie of Tohoku University and David Wright of Oklahoma State University had made some progress towards an answer in 1992 but were, in retrospect, missing a key element of the Bhargava approach—looking at integer points rather than rational points in the relevant vector spaces. Bhargava has now completely solved this problem for quartic fields and quintic fields, and published a "heuristic" formula for the average number of fields of discriminant d (in other words, an educated guess that cannot be formally proved yet) that works for fields of any degree. For number theorists, results like these—*bona fide* general theorems about quartic and higher fields—are like the first glimpse of a new continent.

Beyond the simple enumeration of fields, there are more subtle questions about the average size of ideal class groups or the average number of particular types of ideal classes, for which mathematicians have published guesses, called "Cohen-Lenstra heuristics." The guesses agree very well with computer calculations, but number theorists had no known way to prove them. Bhargava has now verified some of these guesses. Others are still too hard, not because the strategy of using lattice points in vector spaces is flawed but because the regions in which the lattice points must be counted are so complicated. "The reason counting points is hard is that the regions are not compact," Bhargava says. "They have tentacles. In a compact region, the number of lattice points is basically the volume. But if the region has tentacles with finite volume, they could contain infinitely many lattice points or none at all. Dealing with the tentacles is what was hard in the quintic case. I had to work out what kinds of points are in each of the tentacles, and eventually had to divide the tentacles into 160 different pieces and deal with each one individually."

Are there more boxes in Bhargava's future? Oh, yes. The parametrization of ideal classes in cubic rings already involves $2 \times 3 \times 3$ boxes. Bhargava says that more boxes are on the way, including higher-dimensional ones, but he is not ready to talk about them yet because he has collaborators and students who are working on them. Beyond that, he is looking for ways to bring in even more exotic vector spaces, such as octonions—a non-commutative, non-associative algebra of eight-dimensional "complex numbers." "There was a time when I would have thought, how could octonions have anything to do with number theory?" Bhargava says. "Now I'm a complete convert." Judging from his track record, you can expect that the rest of the number-theory world will soon be converted, too.

> **For number theorists, results like these—*bona fide* general theorems about quartic and higher fields—are like the first glimpse of a new continent.**

Figure 1. *Vortices are one of the most characteristic features of nonlinear fluid flow, and range from ephemeral to very long-lasting. Jupiter's Great Red Spot is a vortex in Jupiter's atmosphere that has persisted for hundreds of years. It is occasionally accompanied by more short-lived storms, as seen in this photograph.*

Vortices in the Navier-Stokes Equations

Barry Cipra

FLUID FLOW CONTINUES TO BE one of the great mysteries of modern physics. The way smoke curls upward from the wick of a blown-out candle, or river water roils around a boulder—these familiar, everyday phenomena are, in many ways, more difficult to understand than quantum mechanics or special relativity. Quantum wave functions, in spite of their abstruseness, satisfy mathematically tractable linear equations. On the other hand, water waves are a truly nonlinear phenomenon. One of the consequences of this nonlinearity is that three-dimensional fluids develop turbulence, and at present there is no mathematical theory that adequately explains all the details of turbulent flow. (See Figure 1.)

Thierry Gallay. *(Photo courtesy of Thierry Gallay.)*

The two-dimensional theory is in much better shape, thanks in part to some recent work by Thierry Gallay at the University of Grenoble and Eugene Wayne at Boston University. Researchers have long known that 2-D flows are not plagued by turbulence, no matter how energetic they are. Gallay and Wayne have now nailed down precisely what happens to 2-D fluids: In the long run, the flow settles down to a well-understood circulating pattern known as an Oseen vortex.

Real fluids are, of course, inherently three-dimensional. The 2-D theory has applications in the study of thin films, such as those in liquid crystal displays, but its main use is as a testbed for techniques that apply more generally—literally, a mathematical "proving ground." Gallay and Wayne's work bears this out. They have extended their analyses to an important class of 3-D flows called Burgers vortices, which have come to be called the "sinews of turbulence."

One of the frustrating features of fluid flow is that the *equations* for it are well known. It's the *solutions* that prove elusive. Fundamentally, all fluids, from the most tenuous gas to the stickiest molasses, obey a set of equations known as the Navier–Stokes equations. Initially derived by Claude-Louis Navier in 1822 and George Gabriel Stokes in 1845, the Navier–Stokes equations do for fluids what Newton's laws of motion do for solid objects: They describe the way fluid masses accelerate due to forces such as pressure.

In the elegant notation of vector calculus, the Navier–Stokes equations are written as

$$\frac{\partial u}{\partial t} + (u \cdot \nabla)u = \nu \Delta u - \frac{1}{\rho}\nabla p, \qquad \nabla \cdot u = 0.$$

where $u(x,t)$ is the velocity field—that is, the function that specifies how fast the fluid is flowing, and in what direction, at each point in space x and each moment in time t—ν denotes

Eugene Wayne. *(Photo courtesy of Eugene Wayne.)*

The 3-D theory leaves a lot to be desired—a million dollars worth, to be precise.

viscosity, ρ is the (constant) density of the fluid, and $p(x, t)$ is pressure. Roughly speaking, the first equation expresses the conservation of momentum; it says that velocity changes only when force is applied. The second equation is even simpler to interpret: It expresses the conservation of mass. In effect it says that any fluid entering one side of an imaginary box is matched by an equal amount of fluid leaving the other sides. (To be precise, this is conservation of mass for an *incompressible* fluid, such as water, for which the density is the same everywhere. For compressible fluids, such as gases, the second equation is $\nabla \cdot (\rho u) = -\partial \rho / \partial t$, where ρ is now a function of x and t. This just says that if there is a net outflow of fluid from an imaginary box, then the density of what remains must decrease by a corresponding amount.)

The difficulty of the Navier–Stokes equations stems from the nonlinear term, in which the components of u are multiplied by their own derivatives—the term $(u \cdot \nabla)u$. Physicists call this the "advection" term because it expresses the fact that momentum can be transported or "advected" from place to place within a fluid. The most obvious physical example of advection is the downstream motion of a whirlpool. (Insects that walk on water actually use this phenomenon to move across the surface, as described in "Fluid Dynamics Explains Mysteries of Insect Motion," p. 87. The insect creates a vortex with its feet, and as the vortex carries momentum backward, the insect is propelled forward.)

The nonlinearity makes it impossible, in general, to find exact, analytic solutions to the Navier-Stokes equations, except for very special initial conditions, such as smooth flow along a pipe. (In dimension one—for incompressible flow—the nonlinearity drops out, but what remains is trivial: The flow is constant in space and only speeds up or slows down depending on changes in the pressure differential.) This means that solving the Navier–Stokes equations in practical settings ultimately depends on numerical methods.

Not that theory becomes irrelevant. If anything, rigorous results are more important than ever. Numerical methods don't give exact solutions, they give *approximate* solutions. Knowing how good an approximation is—or setting up a numerical method to guarantee an approximation within a specified amount of error—requires theoretical analysis.

The 3-D theory leaves a lot to be desired—a million dollars worth, to be precise. Among the unknowns about 3-D fluid dynamics is whether all smooth flows will remain smooth for all time. Put differently, is it possible for a smoothly flowing fluid to develop a discontinuity in a finite amount of time? The Clay Mathematics Institute, a privately funded organization supporting mathematical research, has included this question in its list of seven "millennium challenge" problems, each worth a million dollars (see "Think and Grow Rich," *What's Happening in the Mathematical Sciences*, Volume 5, p. 77).

"Smooth" is actually a technical term entailing a number of analytic conditions. One of these is that the total energy of the fluid, defined as the integral over space of the square of the velocity, be finite. This requires that the fluid be essentially motionless at distances far from the origin. That does not in itself, however, forbid it from developing fine-scale peculiarities everywhere.

It is known that smooth flows (in three dimensions) necessarily remain smooth for at least a short time. It's also known that a sufficiently gentle smooth flow (meaning one with little energy) will remain smooth (and gentle) for all time. Where the cutoff lies, if indeed there is one, is unknown.

As mentioned above, one of the most characteristic features of fluid flow is a property called vorticity. Roughly speaking, a vortex is a swirl of fluid around an axis. A tornado is a vortex, as is bathwater draining from a tub. More technically, the vorticity of a fluid at a particular point in space is the angular velocity of the fluid at that point. Imagine a miniscule blob of fluid around the point x suddenly freezing at time t, and the rest of the fluid disappearing. The tiny ice cube will fly off in space with velocity $u(x, t)$. But, like a golf ball or baseball, it will also have some spin—it rotates around an axis at some rate, either fast or slow (or zero).

Mathematically, vorticity is the "curl" of the velocity field, $\omega = \nabla \times u$. In three dimensions, the vorticity is a vector field: It has a direction and a magnitude at each point in space. Because of this, every point with nonzero vorticity belongs to a "vortex line"—a curve whose tangent at each point is parallel to the vorticity at that point. In a minor miracle of vector calculus (having nothing to do with the Navier–Stokes equations—it's true for the curl of any vector field), the *magnitude* of the vorticity is constant along each vortex line.

For a fluid with no viscosity—what's colorfully called "dry water"—vortex lines move with the fluid. That is, if two fluid particles lie on the same vortex line initially, they continue to lie on the same vortex line. (The magnitude of the vorticity may change in time, though.) When v is positive, however, the vorticity diffuses: Particles that start nearby on the same vortex line drift onto increasingly remote vortex lines. This explains why smoke rings grow. Three-dimensional vorticity is extremely complicated, and is closely associated with turbulence.

2-D vorticity is much simpler, mainly because there are no vortex lines. For fluid flow in, say the xy plane, all rotations take place around axes that point in the "z" direction (i.e., perpendicular to the plane). Consequently, the curl of a 2-D vector field is a scalar field—the vorticity at a point in the plane is a number, just as pressure (or density, temperature, and any of a thousand other things) is a number.

The 2-D Navier–Stokes equations can be rewritten in terms of vorticity:

$$\frac{\partial \omega}{\partial t} + (u \cdot \nabla)\omega = v\Delta\omega.$$

There is one especially simple solution to this equation, known as an Oseen vortex (see Figure 2, next page). In this solution, the fluid rotates around the origin in the xy plane at a rate that decays in time and space. At any given moment in time, the vorticity profile is a classic, bell-shaped curve; as time increases, the bell shape spreads and flattens out. The Oseen vortex can be thought of as a (two-dimensional) tornado that slowly spins down to an imperceptible, circulating zephyr.

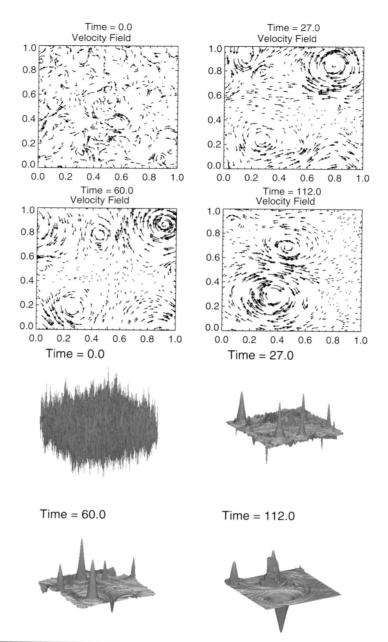

Figure 2. *This figure shows four snapshots in time of the vorticity function of a 2-dimensional fluid evolving according to the Navier-Stokes equation. The top figure is a representation of the velocity field of the fluid in two dimensions. The bottom figure is a graph of the vorticity function. This is not a literal picture of the fluid (which remains confined to a plane); however, it is possibly easier to grasp intuitively, with the downward dimples corresponding to places where vorticity is negative and the fluid is rotating clockwise, and the upward bumps corresponding to points where vorticity is positive and the fluid is rotating counterclockwise. (Figures are based on the work published in W. H. Matthaeus, W. T. Stribling, D. Martinez, S. Oughton and D. Montgomery. "Selective Decay and Coherent Vortices in Two-Dimensional Incompressible Turbulence," Phys. Rev. Lett., **66**, p. 2731 (1991).*

Gallay and Wayne have shown that *every* solution of the 2-D Navier–Stokes equations, no matter how complicated it starts out, enjoys (or suffers) the same fate as the Oseen vortex. More precisely, as long as the initial vorticity is integrable ($\int |\omega(x,0)|\,dx < \infty$), then, as time goes to infinity, the vorticity converges to the bell-shaped profile of the Oseen vortex. Gallay and Wayne were also able to get precise bounds on the *rate* of convergence.

Earlier researchers had proved convergence to the Oseen vortex in restricted settings, such as when the total circulation (that is, the integral of vorticity over the entire plane) is sufficiently small. This can be likened to theorems showing that laminar flows, such as water flowing in a pipe, are stable as long as the flow is not too fast, or, as mentioned earlier, that sufficiently gentle smooth flows in three dimensions remain smooth for all time. Gallay and Wayne have removed these restrictions. Their approach uses techniques from the theory of dynamical systems, which more commonly deals with systems of ordinary, rather than partial, differential equations.

The main result also accords with experimental and numerical observations of 2-D flow. These observations indicate that, even when the flow begins with lots of small-scale eddies, tiny vortices tend to coalesce, forming larger and larger patterns of circulation—a phenomenon known as an inverse energy cascade. Gallay and Wayne's theorem says this process ultimately ends with the fluid rotating as a single vortex. In essence, the Oseen vortex is the only stable flow pattern for 2-D fluids.

The situation is exactly the opposite in three dimensions. The theory of turbulence implies an energy cascade in which big eddies break up into smaller and smaller eddies, down to the scale at which viscosity dissipates the energy into heat. (In the special case of "dry water" there is no viscosity, and the cascade never ceases.) Stability is presumably a property only of flows that are sufficiently gentle to prevent the cascade from getting started.

Nevertheless, Gallay and Wayne have succeeded in extending some of their analysis to three dimensions. In particular, they have obtained stability results for solutions to the 3-D Navier–Stokes equations known as Burgers vortices (named for the twentieth-century Dutch physicist Johannes Burgers). These vortices can be pithily characterized by the description "inward and upward, spinning as you go": A particle immersed in a Burgers vortex (see Figure 3, next page), starting say at (x, y, z), spirals inward toward the z axis while moving faster and faster away from the origin (unless it starts in the xy plane, in which case it simply spirals inward to the origin—that is, the velocity in the z direction is proportional to z).

If the inward impetus toward the z axis is the same from all directions—what's called the symmetric Burgers vortex—the solution is essentially a three-dimensional extension of the bell-shaped Oseen vortex: The vorticity profile is constant as a function of z and bell-shaped in x and y. But again there is a big difference between two and three dimensions. Unlike the two-dimensional Oseen vortex, the three-dimensional Burgers vortex refuses to die out. It just keeps on going for all time, like a tornado that stays over one spot. This is called a "stationary" solution, although the name is a bit misleading: the fluid

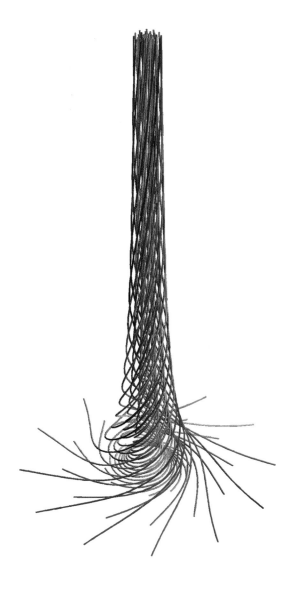

Figure 3. *The Burgers vortex, a three-dimensional version of the Oseen vortex, behaves quite differently. Assuming a limitless supply of fluid coming in "from infinity," the vortex will remain steady for all time (perhaps an analogue to Jupiter's Great Red Spot?). Gallay and Wayne have now proved the existence of Burgers vortices with an elliptical shape, which will remain steady for all time and which are not disturbed by slight perturbations in the vorticity function (at most, the vortex will drift to a different location). (Graphic created by Michael Trott using Mathematica.)*

is moving, but the way in which it moves does not change over time. (A way to understand why this might happen is that the continuing inrush of fluid supplies energy to keep the vortex alive.)

In 1984, Allen Robinson and Philip Saffman at the California Institute of Technology showed that the symmetric Burgers vortex is stable against small perturbations in vorticity, as long as the total circulation is sufficiently small and, more importantly, the vorticity remains oriented along the z axis (which means the circulation is essentially two-dimensional). They also gave an analytic, but non-rigorous, argument for the existence of stationary solutions even when the inrush of fluid toward the axis is not the same from all directions, so that the spirals are elliptical rather than circular. This was especially important because all experimental evidence, both laboratory and numerical, indicates that non-symmetric vortices are the norm.

Gallay and Wayne have given a rigorous—and, it turns out, fairly simple—proof for the existence of non-symmetric Burgers vortices. They have also shown that these stationary solutions are stable "with a shift" against small perturbations in vorticity, even if the perturbation has a component in the z direction. "With a shift" means that a slightly perturbed non-symmetric Burgers vortex will settle back down again to a vortex, but not necessarily to the original one. Some of their results require the total circulation to be small, others require the vorticities to be nearly symmetric. It is unclear whether these restrictions are necessary.

These results do not, of course, answer all the questions about 3-D fluid flow—or even come anywhere close. The nature of turbulence and its relationship to the Navier–Stokes equations are likely to remain open problems for decades, if not centuries, to come. But each analytic advance, while just a drop in the bucket, brings researchers that much closer to a full understanding.

> **But each analytic advance, while just a drop in the bucket, brings researchers that much closer to a full understanding.**

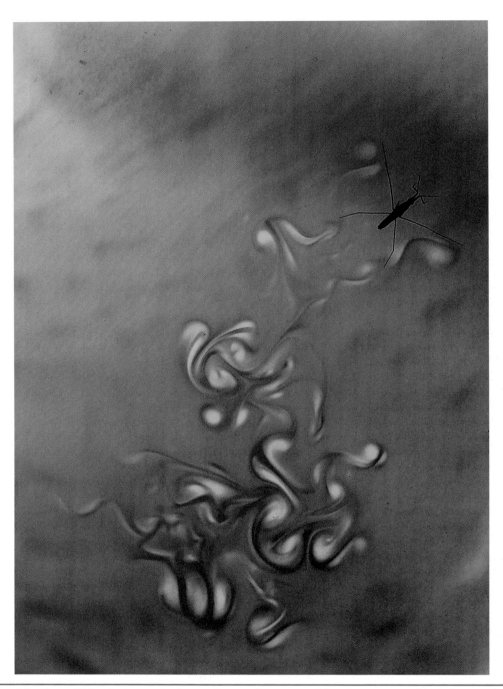

Trailing Vortices. *This photograph shows the vortices created by the water strider's legs as it moves across the water. Vortices are made visible by thymol blue. Bush has shown mathematically, and documented with photographs, that the vortices carry momentum backwards. By conservation of momentum, this makes the water strider move forward. (Photo courtesy of John W. M. Bush.)*

Fluid Dynamics Explains Mysteries of Insect Motion

Dana Mackenzie

WANTED: *Three miniature vehicles. Vehicle 1 should be able to hover in midair without going forward. It should not use a helicopter-like motion, but should instead beat its wings in an inclined plane, generating four times as much upward force as conventional aerodynamics predicts.*

Vehicle 2 should be able to walk on water. In fact, it should be able to sprint and jump when conditions demand. It should also be able to climb a frictionless slope, and do so without moving any part of its body.

Vehicle 3 should be able to push itself through a highly viscous fluid and "fly" through a thinner fluid, and, moreover, it should be able to sense the difference between the two and adjust its method of locomotion accordingly.

All three vehicles should range in size from a few millimeters to a few centimeters. The winning designs should be backed up by at least 100 million years of field testing. Please submit proposals to . . .

John Bush. *John Bush (center) with graduate students David Hu (left) and Brian Chan (right). (Photo courtesy of John W. M. Bush.)*

No, this isn't a request for proposals from a real government agency, but the three "vehicles" described here do exist. Vehicle 1 is a dragonfly, Vehicle 2 is a water strider, and Vehicle 3 is an Antarctic mollusk known as the "sea butterfly." Together, their literally superhuman abilities show the vast differences between life at the scale of insects and life at a human scale.

"The world of water striders is dominated by surface tension. We live in a world dominated by gravity, so we have very poor intuition where surface tension is concerned," says John Bush, an applied mathematician at the Massachusetts Institute of Technology who has studied the motion of water-walking insects for about five years. Human intuition also fails to understand the hovering flight of dragonflies, which toss invisible vortices of air off with their wings as if they were tennis balls. We struggle also to imagine what life is like for the sea butterfly, which grows wings and uses them to "fly" through the water.

Bush and other applied mathematicians and physicists, such as Steve Childress of New York University and Jane Wang of Cornell University, are using mathematical methods (along with careful observation) to learn some of the secrets of the insects. Though his interest is purely fundamental, Bush says that engineers are beginning to take these lessons very seriously, as they try to design micro-air vehicles and microfluidic devices that will use the same mechanisms and operate on the same scales as the insects.

Hitting Their Stride

As you can see by taking a trip to a pond on a summer day, a water strider is a six-legged critter with a body about 1 to 4 centimeters in length and gangly legs that can make its total length closer to 20 centimeters. The legs are bent in such a way that a long piece, called the "tarsal segment," rests on the surface of the water. It acts like a snowshoe that distributes the bug's weight over a larger area and prevents it from breaking through the surface of the water. (See Figure 1.)

There is no great mystery about what holds the water strider up. As Bush said, it is the curvature force resulting from surface tension, which makes the surface behave like a trampoline. The strider's feet make little dents in the surface of the water, but as long as the force exerted by any foot does not exceed 140 dynes per centimeter, it will not and cannot break through. A quick calculation shows that, with a weight of 10 dynes distributed over a total tarsal length of about 10 centimeters, the water strider has a large margin of safety.

Figure 1. *An adult water strider, Gerris remidis. It is easy to understand how a water strider stands on the water: surface tension from the dimples in the water creates an upward force on the animal. However, until recently scientists have not been able to explain how the strider moves on the water. (Photo courtesy of John W. M. Bush.)*

But *moving* on water is another problem entirely. How can you walk on a surface that is practically frictionless? Why don't water striders slip and flail around like humans walking on a patch of ice? For a while, biologists thought they knew the answer. The same surface tension forces that hold the water strider up also provide a little bit of resistance to its movements. When the strider pushes against the water, it creates a little packet of "capillary waves" that move backwards and transport momentum with them. According to Newton's Law of conservation of momentum, if the wave momentum is carried backward, then the insect must move forward.

But there is a problem with this explanation, first pointed out by Mark Denny of Stanford University in 1994. Capillary waves in water have a minimum speed of 23.2 centimeters per second, which can be computed from the density and surface tension. But infant water striders, as shown in laboratory experiments, cannot move their legs that fast. That means they cannot generate capillary waves. No capillary waves means no momentum transfer and no fun for the juvenile water strider.

When he first read about Denny's paradox, Bush immediately saw one flaw in the argument: biologists were assuming the strider's leg traveled at a steady rate. It's true that a paddle traveling through water at less than 23 centimeters per second will not generate a capillary wave, Bush says. But the leg motion is not a steady straight-line motion; it is a short impulse followed by a return stroke. The theory of steady motions does not apply.

To see what was really happening, Bush's graduate student David Hu went out to Walden Pond to collect some water striders, bring them back and watch them walk. "From the very first day in the laboratory, it was clear that they were shedding vortices," says Bush. Vortices are a hallmark of unsteady fluid flow. Engineers and weather forecasters may hate them, but insects love them—and water striders wouldn't be able to go anywhere without them. The discovery of the vortices did not, of course, refute Newton's Law. The water strider still moves by means of momentum transfer, but it is the vortices, not the capillary waves, that do the lion's share of the work.

Eventually, Bush and Hu used a chemical dye called thymol blue to make gorgeous images of the vortices trailing away behind a moving water strider, just like the footprints of a rabbit through the snow. (See Figure "Trailing Vortices," p. 86.) Another student, Brian Chan, developed a robotic water strider, an ingenious contraption that Bush describes as "truly scientific research on a shoestring." The strider was powered by a piece of elastic thread taken from Chan's sock! Research groups at Carnegie-Mellon and Columbia University are now working on more sophisticated versions of Chan's Robostrider, with more powerful energy sources such as a solar cell. (See Figure 2, next page.)

The striders' ability to walk on water is but one of many amazing feats performed by water-walking insects. How do

Vortices are a hallmark of unsteady fluid flow. Engineers and weather forecasters may hate them, but insects love them—and water striders wouldn't be able to go anywhere without them.

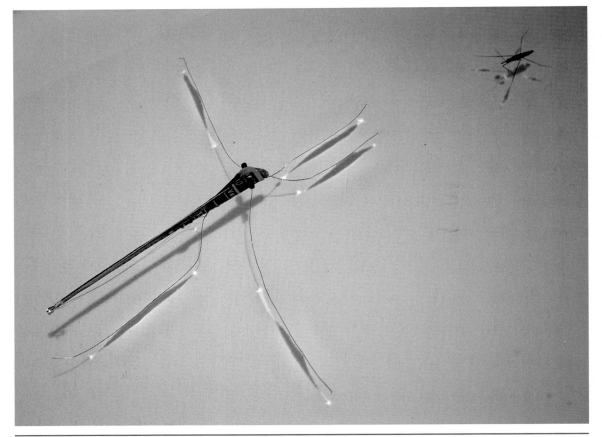

Figure 2. *Chan created a mechanical water strider, powered by an elastic thread from his shoelace, that could take a few steps before running out of power. Other scientists have since developed more sophisticated versions. (Photo courtesy of John W. M. Bush.)*

such insects get out of the water? As Figure 3 shows, an insect perceives the meniscus at water's edge very differently from us. To the insect, it is a frictionless hill that is taller than the insect itself, sloping up to a daunting angle of 40 degrees. If you imagine trying to climb a 40-degree slope of ice that is several feet high, you can begin to appreciate the water strider's problem.

Relatively large water-walking insects cheat by jumping to the top of the hill. Those that cannot jump apply a far more ingenious solution, which also relies on surface tension. For example, an insect called the water treader (*Mesovelia*) has miniature claws that are hydrophilic (water-attracting)—unlike the rest of the leg, which is hydrophobic (water-repelling). The treader plucks the water's surface up a little bit with its front claws, while pressing down with its middle claws. It holds this fixed posture, and in no more than a tenth of a second (so fast that it can only be seen with high-speed photography) it *accelerates* to the top of the hill!

What makes this trick work? It is the same principle that makes corn flakes stick together in a bowl of milk, or bubbles stick together in a glass of champagne. Whenever two menisci get close enough to each other, they tend to attract. The meniscus on one side and the insect's claws on the other side bend

Figure 3. *A water treader, Mesovelia, faces a seemingly impossible task. How can it climb a perfectly frictionless surface, slanted at an angle of 40 degrees, in order to get out of the water? Cheating (by grabbing the wall of the container) is not allowed. (Photo courtesy of John W. M. Bush.)*

the surface layer of the water into the shape of a "U". The water always tries to minimize its surface energy, and it does this by reducing its area. Thus a force is generated that pulls the two sides of the "U" together until they coincide. The force is quite a strong one—stronger than the gravitational force on the water strider—and so the whole thing is over in less than the blink of an eye.

Water-walking insects are not the only animals that use the meniscus trick. When a waterlily leaf beetle falls into the water, it curls its tail upward and lifts the surface of the water. It too glides up the meniscus in a fraction of a second. "They always go tail first, and I wonder if they do it to avoid concussions," Bush says, perhaps only partly in jest. (See Figure 4, next page).

Bush is now studying capillary feeding, which has been observed in birds: How can a bird suck a liquid up into its beak, when it can't pucker its lips to create suction? Some birds use surface tension: they draw a drop into the tip of their beaks; then, in its quest to minimize surface energy, the drop is drawn towards the bird's mouth. Bush believes that some insects may use similar capillary feeding techniques, but no one has observed them because it's too hard to see. Where surface tension is concerned, humans have a lot to learn from animals. "I love working in this area because you gradually come to the conclusion that any mechanism you can imagine that works is already out there," Bush says.

A Sea Change

Steve Childress is another mathematician who doesn't mind traveling a small distance to study fluid dynamics. In his case, the research took him to Antarctica, where he went with biologist Robert Dudley of the University of California, Berkeley, to study a small mollusk called *Clione antarctica*, or the sea butterfly.

When they got to McMurdo Station in November 2000, Childress and Dudley seemed to be out of luck. A cold spring had wiped out the anticipated "spring bloom" of sea butterflies.

Figure 4. *A waterlily leaf beetle, Pyrrhalta, demonstrates the solution to the meniscus-climbing problem. By arching its back, it creates curvature in the surface of the water. Surface tension then creates a strong horizontal force that allows it to reach the edge of the water even though it has to go "uphill." The water treader uses the same solution, except that it plucks the water surface upward with its forelegs and hind legs, while pushing down with the middle legs. (Photo courtesy of John W. M. Bush.)*

Childress and Dudley could not find a single adult—only juveniles. However, this disappointment actually was a stroke of serendipity because it forced them to pay serious attention to the juveniles for the first time.

Adult sea butterflies are about 1.5 centimeters long, but the juveniles are only 3 to 4 millimeters. They have a cigar-shaped body with stubby "wings," and, in addition, they have three bands of cilia roughly circling their head, waist, and tail. The cilia pose an interesting question. Why does one creature need two ways to get around?

In laboratory experiments, Dudley changed the viscosity of the water by adding chemicals to it, and observed how the sea butterfly larvae responded. When the water was more viscous, they would use their cilia to swim, and tuck their wings into their bodies. When the viscosity was reduced, out came the wings. "Dudley could almost train those creatures to stick out their wings and flap them," Childress says. Childress realized that what he was seeing was an adaptation to moving through two different kinds of fluid. Only in the natural setting, it isn't the viscosity of the water that changes—it's the size of the animal as it grows.

Intuitively, the most important quantity for describing a fluid's flow is its viscosity. A viscous or "thick" fluid, such as molasses, tends to flow along smooth streamlines, with no vortices or turbulence. An inviscid or "thin" fluid, such as air, flows in a more complicated way, with lots of turbulence. But for a body moving through a fluid, the size and velocity of the body also play an important role. If the body is very small or moving very slowly, it will not generate eddies even in a thin fluid. Thus it is a combination of factors—the viscosity v, the velocity u, and the length of the body L, that governs the fluid flow. These factors can be summed up in a combined parameter called the Reynolds number:

$$Re = uL/v.$$

Thus, a 2-millimeter animal, moving through water, will experience the same Reynolds number as a 2-meter human moving at the same speed through a fluid that is 1,000 times more viscous than water—a very thick soup indeed!

In the regime of low Reynolds numbers, a well-known theorem of fluid mechanics, called the scallop theorem, says that you cannot propel yourself forward with any sequence of body configurations that is reversible in time. (Thus, wing-flapping or dolphin-kicking are ineffective.) However, it is possible to make forward progress with an asymmetric motion. For instance, you can row: press against the water with a flat paddle, then turn the paddle 90 degrees and move it back to its starting point in a way that minimizes drag. Most sub-millimeter sized animals use a different method. They have cilia or flagella that propel them through the water with a corkscrew motion. This is consistent with the scallop theorem because the time reversal of a left-hand screw is a right-hand screw, which is distinguishable from the original motion.

At high Reynolds numbers (say, Re $>$ 1,000), a whole different set of principles comes into play. This is the realm of bird flight, and at even higher Reynolds numbers (Re $>$ 1,000,000), airplane flight. Mathematicians understand high Reynolds numbers very well. Birds propel themselves in a manner reminiscent of water striders, by flapping their wings and generating vortices of air. As they push the vortices backwards, they receive an equal amount of forward momentum. Thus the main purpose of flapping is to provide thrust. In an airplane the thrust is produced in other ways—for example, by propellers—and the function of the wings is to provide lift (for which purpose a fixed wing is sufficient.)

Most animals, though, live in the intermediate zone between low and high Reynolds numbers, where the equations of fluid mechanics, called the Navier-Stokes equations (see "Vortices in the Navier-Stokes Equations," p. 78) are not as easy to analyze. The sea butterfly juveniles start at a Reynolds number of 10 and grow to a Reynolds number of 100 or so. Through Dudley's experiments, Childress discovered that the sea butterflies' swimming speed while flapping slowed down at the lower Reynolds number, and (when extrapolated) vanished at a value of about 12. This indicated a finite, nonzero threshold

> **Birds propel themselves in a manner reminiscent of water striders, by flapping their wings and generating vortices of air.**

for flapping flight. As the Reynolds number increases, flapping becomes more and more efficient.

It appears that sea butterflies have discovered a theorem that mathematicians were not aware of. The scallop theorem (which says that flapping has zero efficiency) has been proved only at a Reynolds number of 0, but the sea butterfly data strongly suggests that a bifurcation takes place around Reynolds number 12. Below this number, a flapper may jiggle around randomly but it won't be able to move. It is trying to create vortices, but the vortices diffuse away before they can generate thrust. Above this number, a spontaneous symmetry-breaking occurs. Any little push on the animal will create a fore-aft asymmetry, and the fluid is now thin enough that the animal can exploit it. When it flaps its wings, the fore-to-aft fluid flow will carry the eddies away before they disperse. The animal can now fly. (See Figure 5.)

Figure 5. *(a) At low Reynolds number, a flapping wing cannot generate propulsion because it generates symmetric fore and aft vortices. (b) However, at a certain threshold value of the Reynolds number, the symmetry breaks spontaneously. (c) Above the threshold value, flapping flight becomes feasible. Note that the mechanism of propulsion is the same as that of the water strider: flapping creates vortices that carry momentum backward, allowing the animal to move forward. (Figure courtesy of the Applied Mathematics Laboratory, Courant Institute of Mathematical Sciences.)*

So far, no human mathematician has proved the "sea butterfly theorem" yet. Childress verified it for a simplified version of the Navier-Stokes equations, called the Oseen model (see "Vortices in the Navier-Stokes Equations," p. 78), which ignores the feedback between a moving body and the surrounding fluid. (In the Oseen model the flow of the fluid affects the motion of the body, but not vice versa.) For a flapper based on the shape of the sea butterfly, he estimated that the bifurcation occurs at Reynolds number 36. Due to the simplifying assumptions of the Oseen model, that number is clearly an overestimate, but it is a proof of principle that the bifurcation exists. "Nature has probably smoothed out the dividing line by being clever, but what we've done is make it very precise," Childress says.

Dragonflies and Falling Paper

In the late 1990s, when she was a NSF-NATO Postdoctoral Fellow in physics at Oxford University, Jane Wang went to the library to look for a book on random matrices. What she found instead was a new direction for her career. She happened to pick up a book by Childress, called *Mechanics of Swimming and Flying*. "What a fascinating thing to study!" she thought. Later she worked on a postdoctoral project with him, and now she has become one of the leading researchers on the mathematics of insect flight. Recently she proposed a new theory of how dragonflies manage to hover in place.

To watch a dragonfly is to be amazed at its ability to stop and start in midair; even other hovering animals, such as hummingbirds, don't seem to have quite the same precision of movement. Staying airborne without moving forward has always been a challenging aeronautical problem. Fixed-wing aircraft and most birds can't do it because their lift results directly from the flow of air past the wing. Helicopters make do with a rotating wing. Hummingbirds, like helicopters, rotate their wings in a mostly horizontal plane, deriving lift from the flow of air past the wings.

The conventional "lift coefficient" of a wing reflects the dependence of a wing's lift on its airspeed. An ideal two-dimensional wing develops a lift L that is proportional to the square of the wing's speed u, the density of air ρ, and the wing's cross-sectional area S. Specifically, the lift is given by the formula:

$$L = \frac{1}{2}C_L\rho u^2 S$$

where C_L, the lift coefficient, equals $2\pi\sin\alpha$, and α is the "angle of attack" of the wing. Thus a horizontal wing (in theory) generates no lift. A wing improves its lift by tilting slightly, so that $\alpha > 0$. However, if it tilts too much, it stalls (which means that the flow of air off the trailing edge is no longer smooth). In practice, this typically occurs around $\alpha = 15°$, so the lift coefficient hits its maximum value somewhere between 1 and 2.

Dragonflies are puzzling to a physicist for two reasons. First, their wings do not move in a nearly horizontal plane, but a steeply inclined one. Second, the first estimates of the lift coefficient of their wings seemed to be about four times too large to be reasonable—between 3.5 and 6. Of course, there are several problems with applying the aerodynamic formulas for aircraft to the wings of a dragonfly. Those formulas assume a steady state—a fixed wing moving at constant speed through a fluid. But a steady-state model doesn't apply to dragonflies, just as it did not apply to John Bush's water striders. The anomalous lift of the dragonfly wings may come in part from their ability to generate vortices and propel them downward. This almost certainly seems to be the case for many insects, from fruit flies to hawkmoths. (See Figure 6, next page, and see Figure 7, p. 97.)

However, Wang found an even more important mistake in the conventional analysis of dragonfly flight. Every wing generates drag as well as lift. Drag is the force parallel to the wing's

To watch a dragonfly is to be amazed at its ability to stop and start in midair; even other hovering animals, such as hummingbirds, don't seem to have quite the same precision of movement.

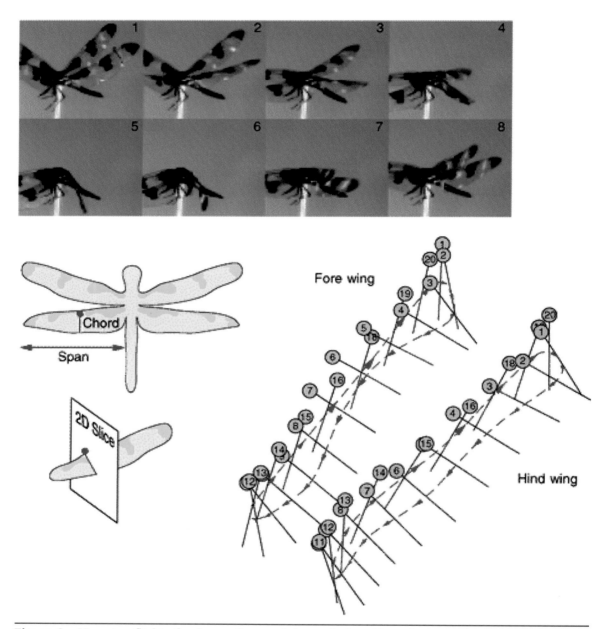

Figure 6. *Hovering flight of a dragonfly posed theoretical problems because the dragonfly wing appears to generate more lift than any known airfoil. In this sequence of pictures, note the steep slope of the downstroke (inclined at 60 degrees), which Jane Wang says is optimal for hovering flight. (Reprinted, with permission, from the Annual Review of Fluid Mechanics, Volume 37, © 2005 by Annual Reviews, www.annualreviews.org.)*

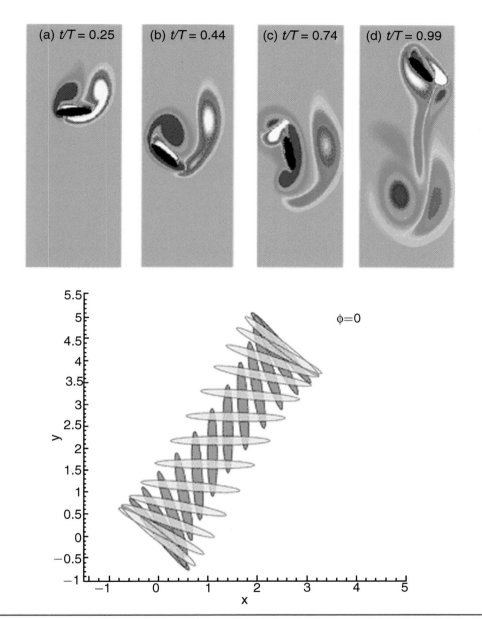

Figure 7. *Like the water strider and the sea butterfly, the dragonfly creates vortices and propels them in the opposite direction from the way it wants to go. (Reprinted, with permission, from the Annual Review of Fluid Mechanics, Volume 37, © 2005 by Annual Reviews, www.annualreviews.org.)*

> **As the wing is moving downward along this path, aerodynamic drag resists its motion—and therefore the drag has a very large upward component. It would be absurd for the dragonfly to minimize drag, because the drag is helping it out!**

motion, while lift is the force perpendicular to the motion. To an aeronautical engineer, drag is always bad because it slows the plane down and contributes nothing to keeping the plane aloft. Engineers always try to keep the lift-to-drag ratio as high as possible. Previous researchers had assumed that the dragonfly wing must also have a high lift-to-drag ratio.

However, that turned out to be incorrect. In high-frequency photos, Wang could see that the dragonfly, when hovering, beats its wings along a steeply inclined path, which makes a 60-degree angle with the horizontal. Because the animal does not have any forward component of velocity, the 60-degree inclined plane marks the true direction of motion of the wing. As the wing is moving downward along this path, aerodynamic drag resists its motion—and therefore the drag has a very large upward component. It would be absurd for the dragonfly to minimize drag, because the drag is helping it out! The dragonfly therefore turns its wing to press against the air like a paddle, creating an angle of attack around 60 degrees and producing lots of drag. On the other hand, during the upstroke, the drag will point downward. On that part of the stroke, the dragonfly does want to minimize drag, so it turns it wing parallel to the airflow to reduce the drag force. (See Figure 8.)

Because the angle of attack of the wing is so large, it is very far into the stalled regime. Any analysis of its motion based on smooth, steady fluid flow is bound to give incorrect answers. So Wang went back to the full Navier-Stokes equations and worked out a model that incorporates both lift and drag. The model showed that 76 percent of the vertical force on the dragonfly's wing comes from drag. In other words, the force is almost exactly 4 times greater than it would be if the wing were using lift alone. This completely resolves the problem of where the dragonfly gets its "extra" lift from. Computer simulations also showed that the wing tosses vortices downward, so that in fact the way a dragonfly stays aloft is very reminiscent of the way a water strider moves forward. Finally, the simulations demonstrated that the vertical force due to drag decreases sharply if the plane of motion of the dragonfly wing gets steeper than 60 degrees, because of the effects of unsteady flow (the interaction between the wing and the flow generated in the previous stroke). So it seems to be no accident that dragonflies beat their wings at that angle.

Wang and her students are currently trying to determine optimum wing-motion patterns for several other kinds of flying animals, including fruit flies, hawkmoths, and bumblebees. So far they are finding that the optimal beating frequencies and wing trajectories are close to the ones that the animals actually use. Once again, nature seems to have figured out the best possible solution long before mathematicians did. However, without mathematics we certainly would not be able to appreciate the efficiency of the solutions that nature has evolved.

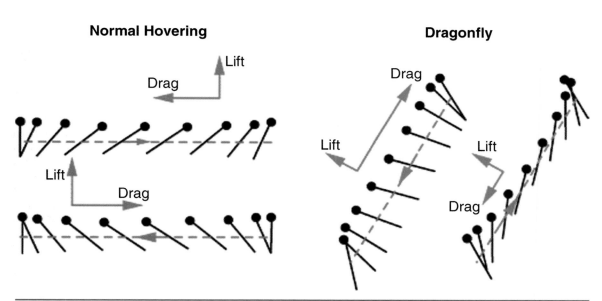

Normal Hovering

Lift

Drag

Lift

Drag

Dragonfly

Drag

Lift

Lift

Drag

Figure 8. *Hummingbird hovering and dragonfly hovering may seem similar, but the physical mechanisms are quite different. Hummingbird wings move in an essentially horizontal plane; thus, the hummingbird stays aloft by lift alone. Dragonfly wings operate in a "highly stalled" regime, which is bad for hummingbirds and airplanes but perfect for dragonflies. They use both lift and drag to stay aloft. Note also that the horizontal orientation of the wing during the downstroke creates maximal drag. On the upstroke, the dragonfly orients its wing vertically to minimize drag. (Reprinted, with permission, from the Annual Review of Fluid Mechanics, Volume 37, © 2005 by Annual Reviews, www.annualreviews.org.)*

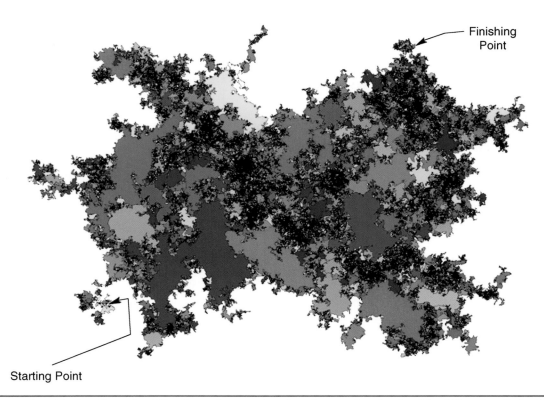

Finishing
Point

Starting Point

Brownian Motion Frontier. *The Brownian motion frontier can be obtained by filling in all the holes formed by a Brownian motion; the frontier is the boundary of the filled-in region. Lawler, Schramm and Werner have proved that this frontier always (that is, with probability 1) has a fractal dimension equal to 4/3. (Graphic created by Michael Trott using Mathematica.)*

Brownian Motion, Phase Transitions, and Conformal Maps

Dana Mackenzie

IN 1827, A SCOTTISH BOTANIST named Robert Brown made a puzzling discovery that would eventually make his name famous—not among botanists, but among physicists. While looking through a microscope at pollen grains suspended in water, he saw them jiggling around as if they were alive. At first he thought they *were* alive, but then he noticed that pollen grains from plants that had been dead for a century showed the same rapid vibrations.

Over the next century, physicists observed this random motion, now called "Brownian motion," in all sorts of non-living particles, and they eventually identified the cause.

The jiggles are the effects of innumerable random collisions of the particles with molecules in the liquid they are suspended in. If we could see individual molecules of gas or liquid, we would see them travel in very much the same sort of trajectory. Brownian motion is ubiquitous in nature.

Classical mathematics, the calculus developed by Isaac Newton, deals only with smooth curves that have a definite tangent at any point (possibly with the exception of finitely many corners). But the trajectories of Brownian motion are so jagged that *every* point is a corner. In 1982, Benoit Mandelbrot brought such curves (which he called "fractals") to wide popular attention with his manifesto, *The Fractal Geometry of Nature.* The book includes some pictures of Brownian motion trajectories and an audacious guess about their "fractal dimension," which is a rough measure of how jagged they are. However, Mandelbrot could not prove his guess, and neither could anyone else for many years.

Beginning in 2000, a group of three probabilists—Greg Lawler, who was then at Duke University, Wendelin Werner of the Université de Paris-Sud, and Oded Schramm of Microsoft Research—finally began to unravel many of the secrets of Brownian motion. "Their work is one of the finest achievements in probability in the last twenty years," says Yuval Peres of the University of California at Berkeley, "and I know this feeling is shared by many probabilists, analysts, and mathematical physicists."

Perhaps most importantly, their methods also apply to many other categories of "random curves," including some types that are very important to physicists. Brownian motion has a property called conformal invariance, which physicists believe is common to all 2-dimensional systems that undergo a

Oded Schramm. *(Photo courtesy of Oded Schramm.)*

Wendelin Werner. *(Photo courtesy of Wendelin Werner.)*

phase transition. (An example is the transition of a ferromagnetic metal from an unmagnetized to a magnetized state). In 2002, Stanislav Smirnov of the Royal Institute of Technology in Stockholm proved rigorously that a certain version of percolation in two dimensions obeys conformal invariance. Thus, percolation became the first phase transition that could be modeled in a mathematically rigorous fashion using the theory developed by Lawler, Werner, and Schramm.

From Random Walks to Brownian Motion

To understand Brownian motion, it helps to begin with a simpler kind of random motion, called a random walk. Imagine that you are in the middle of an unfamiliar city, whose streets form a rectangular grid. You don't have a map, so you set out walking randomly. At each intersection, you flip a coin twice to decide which way to go next: left, right, forward, or back the way you came. (Each choice has probability $1/4$.)

The good news is that, if the city has finite size, you will eventually get to its edge. The bad news is that it will take much longer than the more sensible strategy of walking in a straight line. If a straight line gets you to the edge of the city in n steps, the random-walk strategy will almost surely get you there in roughly n^2 steps. In the process you will have visited a significant proportion of the intersections in the city, some of them many times.

Now imagine doing a random walk on a finer and finer grid, with each "city block" taking proportionately less time. Eventually, the grid will get so small that you can't even see it under a microscope, and your meandering path will look more and more like Brownian motion. In fact, this is one way to define Brownian motion—as the limit of random walks on an "infinitely fine" grid. As mentioned above, a typical Brownian motion trajectory is so jagged that it has no well-defined tangent at any point. Thus the methods of calculus, which assume such a tangent exists, are completely ineffective for studying these curves—or so it would seem. Moreover, the randomness of Brownian motion, which is at the heart of its definition, would seem to defy the mostly deterministic tools of mathematical analysis. (See Figure 1.)

Nevertheless, Brownian motion has some beautiful and very specific properties. It may be random, but it is not without structure. An expert can readily tell a Brownian motion trajectory apart from some other jagged fractal curve that has not been generated by Brownian motion. First, Brownian motion is *rotationally symmetric*. That is somewhat surprising, because it is not true for random walks. Every random walk is drawn on a lattice with two definite axes. Somehow, at the very last moment, when the squares of the lattice shrink down to infinitesimal size, the motion's directional bias disappears as well.

The proof of rotational symmetry is not hard and uses some basic ideas of probability theory. Because the east-west motions of a Brownian motion are governed by what amounts to an infinite series of coin flips, the position of a particle at time t has a probability distribution given by a "bell-shaped curve." In other words, the probability density of being at position x

along the east-west axis at time t is given by

$$p(x, t) = \frac{1}{\sqrt{2\pi t}} e^{-x^2/2t}.$$

The probability density of being at position y along the north-south axis looks exactly the same:

$$p(y, t) = \frac{1}{\sqrt{2\pi t}} e^{-y^2/2t}.$$

Because the motions along the two axes are independent, the probability distributions multiply. The joint probability distribution for being at position (x, y) at time t is the product

$$p(x, y, t) = \frac{1}{\sqrt{2\pi t}} e^{-x^2/2t} \cdot \frac{1}{\sqrt{2\pi t}} e^{-y^2/2t} = \frac{1}{2\pi t} e^{-(x^2+y^2)/2t}.$$

The last expression is rotationally invariant, as one can readily see from a graph. Because the probability density for a point to lie on a given trajectory is unaffected by rotation, the probability distribution for the entire space of Brownian motion trajectories is also invariant. (Note that the *individual curves* are not rotation-invariant. It is the *probability* of observing any particular curve as a Brownian motion trajectory that is invariant.)

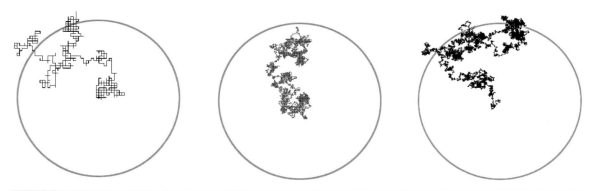

Figure 1. *Random walks on a square grid with n = 1000 steps, 10,000 steps, and 100,000 steps. In each case a circle of radius $\sqrt{n}$ is superimposed, showing that the typical distance traveled in n steps is roughly $\sqrt{n}$. As the mesh size of the grid gets smaller, it becomes harder to tell that the motion was generated by horizontal and vertical steps. In the limit, as the mesh size approaches 0, the trajectory approaches Brownian motion, which is rotation-invariant. Brownian motion is also scale-invariant (a magnified version still looks like Brownian motion). (Figure reprinted from "Scaling limits and super-Brownian motion," by Gordon Slade with illustrations by Bill Casselman, Figure 1, p. 1057, Notices of the American Mathematical Society, October 2002.)*

This is only the tip of the iceberg where invariance properties of Brownian motion are concerned. Brownian motion is also *scale-invariant*, which means that a magnified Brownian motion trajectory still looks exactly the same. (Technically, the time scale must be changed correspondingly. When you zoom in your microscope by a factor of 2, the particle will move out of your field of vision $\sqrt{2}$ times quicker.) Astoundingly, Brownian motion remains invariant even if the magnification factor and the amount of rotation *vary from point to point*. (See Figure 2, next page.) In other words, you can do some rather unspeakable things to your microscope lens—melt it down, swirl it around, massage it like Silly Putty—and as long as the lens still magnifies both directions equally at every point, your Brownian motion will still look like a Brownian motion when you look through the lens.

This very powerful property is called conformal invariance. First established by Paul Levy in 1948, it still does not really have an elementary proof. A nonrigorous explanation would be as follows: If the mesh size of the random walk grid is very small compared to the scale of the distortions in your lens, then in any small region you are looking at the random walk through an essentially normal lens. So in each small region, the Brownian motion limit will still look like a Brownian motion. Unfortunately, this so-called "proof" really has very little to do with the rigorous proof, which involves more sophisticated probabilistic techniques called stochastic integration and Ito's formula.

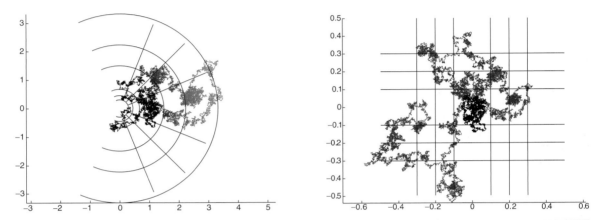

Figure 2. *Brownian motion in the plane satisfies a powerful invariance property, called conformal invariance. Here, the exponential map warps the plane but leaves Brownian motion looking essentially the same. (Figure courtesy of Wendelin Werner.)*

Conforming to Prediction

In their groundbreaking paper of 2000, Lawler, Schramm, and Werner explained how to use the conformal invariance of Brownian motion to make sharp, quantitative predictions. They addressed two types of questions:

- What is the *dimension* of various "interesting sets" connected with Brownian motion? One interesting set is the frontier, or the set of all points that are on the boundary after all loops have been filled in (see Figure, "Brownian Motion Frontier," p. 100). Another interesting set is formed by the "pioneer" points—the points that were on the frontier at the time when they were first drawn (though they may subsequently be enclosed by a loop). A third is the set of "cut points," which would separate the trajectory into two disjoint pieces if a cut were made at that point. For example, every point on a straight line segment is a cut point; no point on a circle is a cut point. The existence of many loops means that most points on a Brownian motion trajectory will not be cut points—but some of them may be.

- Given two Brownian particles on a microscope slide of radius R units, starting 1 unit away from the center, what is the probability that they will hit the edge of the slide without ever having their trajectories cross? The answer depends, of course, on R. The precise form of the dependence gives rise to *intersection exponents* that are closely related to the dimensions mentioned above.

The fractal dimension of a set, mentioned in the first question above, is a description of its scaling properties. When you take a line segment and magnify it by a factor of 2, you get two (2^1) copies of the original line segment. When you magnify a square, its area grows by a factor of four $(= 2^2)$, which means that you have four copies of the original square. But there are some fractals that grow by intermediate amounts. For example, they might grow by a factor $2^{1.6}$ when they are magnified by a factor of 2. Because $2^{1.6}$ is close to 3, what this means is that the twice magnified set looks "essentially the same" as 3 copies of the original set. By analogy with the line segment, which has dimension 1, and the square, which has dimension 2, such a fractal would have dimension 1.6.

Surprisingly, the eye is very good at recognizing fractals of different dimensions. Intuitively, the "rougher" a fractal is, the higher its dimension. A smooth curve (differentiable, in the sense of calculus) has dimension 1. Brownian motion trajectories, which are extremely rough, have dimension 2. The frontier of a Brownian motion is a lot smoother, because it is formed by filling in all the cavities in a Brownian trajectory. In 1982, Benoit Mandelbrot conjectured that the dimension of this outer boundary was 4/3. His guess was based on little more than pictorial evidence from a few computer-drawn images of Brownian frontiers. "A lot of us laughed at him and thought there wasn't enough evidence," Lawler says. But eighteen years later, Lawler helped prove that Mandelbrot was correct.

For good measure, Lawler and his colleagues proved that the set of pioneer points has dimension 7/4 (it is rougher than the boundary itself, because it contains many wiggles that have been erased from the frontier. Also, the cut points have dimension 3/4. A dimension less than 1 means that the cut points form a kind of "dust" that is not quite dense enough to coalesce into a connected curve.

As for the second problem, Lawler, Schramm, and Werner proved that the probability of two Brownian motions escaping the "microscope slide" without ever intersecting each other decreases as $R^{-5/4}$. (Remember that R is the radius of the slide.) In other words, if you make the slide 16 times larger, you will decrease the probability of nonintersection by a factor of 32.

Schramm-Loewner Evolution

The way that the three mathematicians arrived at these results was a fortuitous combination of two independent explorations —or perhaps an intersection of two random walks. For several years, Lawler and Werner had been exploring different varieties of Brownian motion. For example, they were interested in Brownian motion trajectories that avoid intersecting themselves. It is still not clear what is the best way to define

> **Surprisingly, the eye is very good at recognizing fractals of different dimensions. Intuitively, the "rougher" a fractal is, the higher its dimension.**

Of all the ways ever proposed to define a random curve in the plane, SLE must be the least obvious and the most powerful.

a probability distribution on this class of curves. One way, as above, is to consider Brownian frontiers. The self-crossings of a Brownian trajectory always happen in the interior, not at the boundaries, so a Brownian frontier will not have any self-intersections. A second way to avoid intersections would be to approach the Brownian motion with random walks that also avoid themselves. Such walks can be generated, for instance, by taking a standard loopy random walk and erasing all the loops, starting with the biggest ones first. However, the probability distribution of "loop-erase random walks" might not converge to the probability distribution of Brownian frontiers. Indeed, it might not converge to anything at all.

However, Lawler and Werner succeeded in proving that these processes, assuming they were conformally invariant, would satisfy a universality property. Any other process that was conformally invariant and satisfied another property called "locality" would have the same intersection exponents. Thus, you could answer any quantitative question about (say) Brownian frontiers by referring that question to a more convenient member of the same universality class.

At the same time, unbeknownst to Werner and Lawler, Schramm was in the process of discovering the "most convenient member" of this class and many others besides. "Oded did something no one had ever done before, which was to say: suppose I have a limit that is conformally invariant. What do I know about it?" says Lawler. In a *tour de force* of classical analysis and modern probability theory, he showed that any conformally invariant process obeying a "Markov property" (given information about a particle's present, what it does in the future is independent of the past) must have a distribution identical to what he called a "stochastic Loewner evolution." Everyone else now calls this a Schramm-Loewner evolution; conveniently, the acronym (SLE) is the same either way.

Of all the ways ever proposed to define a random curve in the plane, SLE must be the least obvious and the most powerful. Schramm proposed to grow the curve, like a stalagmite growing from the floor of a cave. In this case, the "floor" is the axis of real numbers, and the curve grows upward from there. A deep theorem of complex analysis, called the Riemann Mapping Theorem, says that no matter how convoluted the stalagmite may become, it is possible to find a conformal transformation that translates the exterior of the stalagmite (in other words, all the air in the cave) to the upper half plane (in other words, the cave with the stalagmite removed). The conformal invariance of the process means that the tip of the stalagmite must grow in the same way before the conformal transformation as it does afterwards. The effect of the conformal transformation—and its ingenuity—is that it transforms the problem in such a way that the growth of the curve always occurs on the real axis, the floor of the cave. The random curve can be completely specified by a single real-valued function called the driving function, $W(t)$, which specifies where on the real line the "growing tip" of the transformed stalagmite is at any moment t.

What kind of function is $W(t)$? That is where the probability theory enters in. Schramm showed that, because of the Markov property, only one kind of function will work: a Brownian motion in *one* dimension! There is only one unknown parameter that has to be specified, called the *intensity* of the Brownian motion, which is denoted by kappa (κ). A more intense Brownian motion has larger wiggles.

Combining their results, the three mathematicians could now conclude that any conformally invariant process obeying the Markov property has to be equivalent to a Schramm-Loewner evolution SLE_κ, for some κ. To answer any quantitative question about a process that generates random curves, you only have to prove that it is conformally invariant and then figure out what is the right value of κ. In fact, the result was much more powerful than what Lawler and Werner had originally proved, because it no longer depended on the technical assumption of locality.

New theorems began falling like ripe apples, and they are still falling. The outer boundary of a Brownian motion turns out to be in the class of $SLE_{8/3}$. Computing the dimension of the outer boundary was hard for Brownian motion but easy for $SLE_{8/3}$, because the latter traces out its boundary points more or less in order. The dimension came out to be $4/3$, confirming Mandelbrot's guess of nearly two decades earlier. Loop-erase random walk converges to SLE_2. (Thus, in fact, it is different from the Brownian frontier process.) Schramm and Scott Sheffield proved that a process called the "harmonic explorer" converges to SLE_4. As described below, Stanislav Smirnov proved that a certain type of percolation process converges to SLE_6. In every case, these results lead to corresponding theorems about the dimensions and intersection exponents of the random curves. (See Figure 3, next page.)

Stanislav Smirnov. *(Photo courtesy of Stanislav Smirnov.)*

Phase Transitions

All of these results were of great interest to physicists not so much because of the specific numbers involved—although these, in many cases, confirmed previously published conjectures—but because you could get numbers at all. The intersection exponents strongly resemble the *correlation distances* that are seen in physical systems at a phase transition.

Examples of phase transitions abound in physics. For example, when a ferromagnetic material is heated past a certain point—its Curie point—it loses the ability to become magnetized. The random jittering of its atoms overcomes the tendency of their spins to line up. However, as the temperature is brought down toward the Curie point, some clusters of atoms with parallel spins begin to form. The clusters create a *correlation* between nearby atoms: once you know the spin of one atom, the spin of its neighbors is not completely random. However, the amount of correlation decreases exponentially as a function of distance.

At the Curie point, however, something dramatic happens: infinitely large clusters begin to appear. That means the material is now capable of being magnetized. If you flip the spin of just one atom, the change can propagate all the way out to infinity (or until it reaches the end of your magnet). The correlation of spins now decreases with distance much less rapidly than before. If the atoms are R units apart, it decreases as R^{-d} for some exponent d. In other words, just like the intersection probabilities of Brownian trajectories, the correlation takes the form of a *power law*, with a critical exponent d.

Figure 3. *Schramm-Loewner evolution (SLE) for kappa = 2 (top left), 4 (top right), and 6 (bottom). SLE has turned out to be a model for all sorts of processes that give rise to random curves. The Brownian frontier corresponds to kappa = 8/3, and a critical percolation model corresponds to kappa = 6. (Figures courtesy of Vincent Beffara.)*

Another conjectured similarity between Brownian motion and phase transitions is conformal invariance. For Brownian motion, conformal invariance results from the scaling properties of random walk, plus some very special properties of the "bell-shaped curve." Likewise, physical systems at a phase transition have a built-in scale invariance, because fluctuations at the most microscopic scale ramify up to larger and larger scales, all the way out to infinity. Perhaps with a certain amount of hubris, physicists believe—although they cannot prove it—that the scale invariance extends to conformal invariance, as it does for Brownian motion.

At present, the analogy between Brownian motion and phase transitions is, for the most part, only an analogy. No physical system is known whose phase transition is precisely described by Brownian motion. The intersection of two Brownian trajectories, as Lawler points out, has no physical meaning, because "particles aren't worrying about where they've been." Self-avoiding random walks might be relevant to polymer growth, because a polymer is like a long path that avoids itself. However, nothing rigorous has been proven yet about dimensions of self-avoiding walks.

Even so, the work of Lawler, Schramm, and Werner has motivated physicists and mathematicians to try to prove conformal invariance for physically meaningful models. The first major success came in 2002, when Stanislav Smirnov proved that a particular model of percolation is conformally invariant and converges to SLE_6.

The Fifty Percent Solution

The model that Smirnov has solved is an idealized version of the way water percolates through the soil. For simplicity, we think of a 2-dimensional slice of soil. Suppose that the soil consists of a honeycomb lattice of particle sites, each of which is either empty or full. If a site is empty, then water can flow through it; if it is full, water cannot enter. (See Figure 4.)

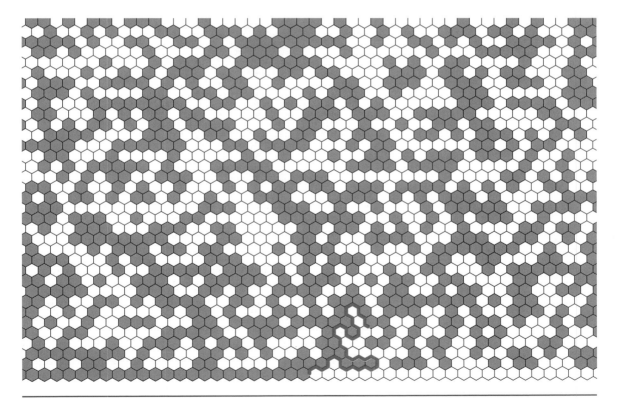

Figure 4. *In the critical percolation model, each cell in a honeycomb has an equal probability of being filled or empty. The boundary between filled and empty regions converges to SLE6, and has recently been shown by Smirnov to be conformally invariant, just like Brownian motion. (Figure courtesy of Wendelin Werner.)*

This 2-dimensional percolation model has a phase transition when the probability, p, of any site being full is 50 percent. When p is greater than 50 percent, the empty cells of the honeycomb form small clusters that are completely surrounded by full cells. Thus water cannot flow through the soil. As p decreases toward 50 percent, the clusters of empty cells grow larger and larger. When p hits exactly 50 percent, so that any site in the honeycomb is equally likely to be empty or full, the clusters of open cells become arbitrarily large. In fact, you can no longer tell whether you have clusters of empty cells surrounded by full ones or clusters of full cells surrounded by empty ones. Finally, as p decreases below 50 percent, the full cells become isolated islands in a network of tunnels, and water can flow freely through the soil.

The percolation model generates a random curve in the following way. If we artificially create an infinite cluster of filled hexagons (all the hexagons at ground level to the left of point zero) and another infinite cluster of empty hexagons (all the hexagons to the right of point zero), then the interface between those two infinite clusters will form a never-ending twisty curve. (See Figure 5.) As in the case of random walk, we then let the size of the hexagons get smaller and smaller. The limit curve looks somewhat like a Brownian trajectory but a little bit less jagged—and that is no accident.

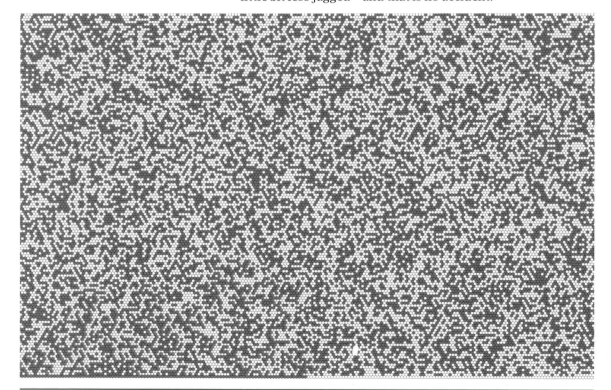

Figure 5. *This is a view of the critical percolation model with a smaller grid size. The model is seeded in such a way that all the honeycomb cells along the negative x-axis are filled and all the ones along the positive x-axis are empty. This guarantees that the interface between the filled region and empty region (heavy line) will never touch the x-axis again, but must continue wandering forever. (Figure courtesy of Wendelin Werner.)*

WHAT'S HAPPENING IN THE MATHEMATICAL SCIENCES

By an ingenious argument that relies very much on the geometry of the hexagonal lattice, Smirnov proved that percolation interfaces are conformally invariant. Therefore, they must be in the same "universality class" as one of the Schramm-Loewner evolution laws, SLE_κ. But which one? As it turned out, there was only one possibility. Smirnov calculated explicitly the probability that water will percolate horizontally through an $m \times n$ rectangle whose top and bottom are all filled (and thus impermeable). (John Cardy, a physicist from Oxford, had conjectured the correct probability several years earlier.) Solving Cardy's conjecture was a little bit like computing an intersection exponent—it was enough to uniquely determine the right value of kappa: $\kappa = 6$. Therefore, percolation interfaces belong to the same class as SLE_6, which has dimension 7/4. This is the same class as the set of pioneer points of Brownian motion.

At present, the hexagonal percolation model is the only phase transition that has been proven to be conformally invariant. Even percolation on a square lattice—seemingly a trivial modification of the hexagonal model—is not yet known to converge to a conformally invariant probability distribution. A variety of other models, such as the Coulomb gas model (for an electrically charged gas) and the Ising model (a simplified version of ferromagnetism) are suspected but not proven to be conformally invariant. Even so, the discovery of Schramm-Loewner evolution was a major step forward. For many physicists, what really matters is not the proof of conformal invariance (which they "know" already in some empirical sense) but the fact that they now have a complete dictionary of conformally invariant random curve models.

Only one inconvenient question remains: what happens in three dimensions? Unfortunately, the recent results apply only to Brownian motion and percolation in two dimensions. Phase transitions certainly happen in the 3-dimensional world, but conformal invariance does not seem to be the right way to think about them. In three dimensions, conformal invariance is a very weak tool, because there are very few conformal transformations of 3-space. More precisely, the group of conformal transformations in 3-space is only finite dimensional, but in 2-space (the plane) there is an infinite-dimensional family of conformal transformations.

At the same time, numerical experiments in 3-dimensional space have not generated any simple predictions for the intersection exponents of Brownian motion. Instead of simple fractions like 4/3 or 5/4, one seems to get mysterious and possibly irrational numbers. As Lawler says, there is also no reason to think that all critical exponents will be governed by a single parameter κ, as they are in 2-dimensional space. Thus the theory in three dimensions remains very primitive. No one knows the right conjectures, and even if we knew the right conjectures, no one would know how to prove them. But you have to learn to crawl before you can fly. Mathematicians have finally mastered Brownian crawling, and perhaps Brownian flight through three dimensions will come in the future.

Mathematicians have finally mastered Brownian crawling, and perhaps Brownian flight through three dimensions will come in the future.

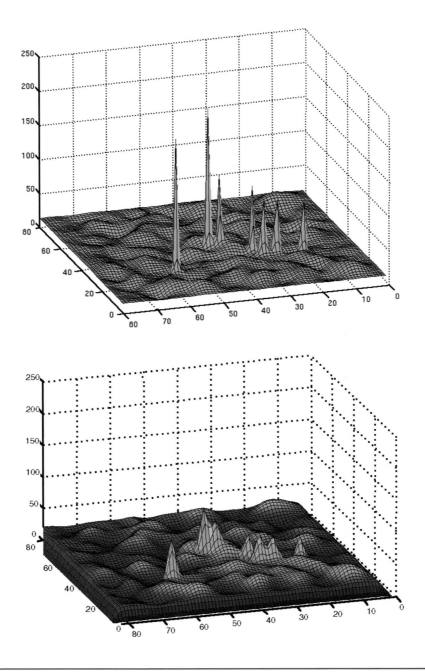

Figure 1. *Many computer algorithms, such as the simplex algorithm, have a "complexity landscape" like the figure at top. Most of the time the algorithm arrives at a solution very quickly. In a very small proportion of cases, shown by the peaks, the run time blows up. Because conventional complexity analysis always assumes a worst-case scenario, it misrepresents the ordinary performance of the algorithm. Smoothed analysis lowers the sharp peaks by allowing small random variations in the input parameters (the coordinates in the "input space" shown here). (Figures courtesy of Daniel A. Spielman.)*

Smooth(ed) Moves

Barry Cipra

W HEN SOMETHING WORKS BETTER IN PRACTICE than it does in theory, there's usually something missing in the theory. That was the case for many years in a branch of mathematics known as linear programming, which is widely used in business applications. Practitioners knew that an algorithm called the simplex method works extremely well on the problems they typically encounter. They also knew the simplex method can be tricked into extremely poor behavior on certain contrived problems. What they didn't know was why such bad behavior never seems to occur in practice.

Now, Dan Spielman at Yale University and Shang-Hua Teng at Boston University have introduced a new way of measuring an algorithm's efficiency that explains why the simplex algorithm works so much better than it is supposed to. Even small random perturbations of the input data for a typical linear programming problem, they showed, suffice to move it away from the regime where the simplex algorithm behaves poorly. Since the data in most real-world problems is at least slightly noisy, it is extremely unlikely the simplex method will ever be presented with a problem it has trouble solving.

A schematic way to understand the result is shown in Figure 1. Here, the two horizontal axes represent different values of input data, and the vertical axis represents the amount of time the simplex method will take to solve the problem, given that initial data. The sharp peaks represent the cases where the simplex method takes a long time. In real life, the initial data will be contaminated by random noise. Spielman and Teng's "smoothed analysis" calculates the likely run time of the simplex algorithm after the random noise is taken into account. As Figure 1 shows, their analysis smooths out the extremely sharp, localized peaks. Even a worst-case scenario no longer looks as bad as it once did—and that is precisely the point.

As originally conceived, smoothed analysis is not a new algorithm: it is simply a new way of measuring the efficiency of existing algorithms. Consequently, it is now being used to understand other algorithms whose typical behavior differs greatly from their worst-case behavior. These include Gaussian elimination (a technique for solving large systems of linear equations, taught in every college linear algebra course), the iterative closest-point algorithm (used in computer vision, to align three-dimensional images), and the k-means method (used in data mining, to separate data points into "clusters" of points that lie close together).

Ironically, Spielman and Jonathan Kelner have recently developed a version of the simplex algorithm that performs relatively quickly even in the worst case—and thus no longer exhibits the defect that smoothed analysis was supposed to remedy. However, Spielman says that they would not have been

> As originally conceived, smoothed analysis is not a new algorithm: it is simply a new way of measuring the efficiency of existing algorithms.

> "This is the sort of result we were hoping smoothed analysis would prompt," Spielman says. "Our hope always was that smoothed analysis would lend insight into why an algorithm worked, and then this insight would lead to a better algorithm."

able to develop this improved algorithm without understanding smoothed analysis first. "This is the sort of result we were hoping smoothed analysis would prompt," Spielman says. "Our hope always was that smoothed analysis would lend insight into why an algorithm worked, and then this insight would lead to a better algorithm."

Algorithmic Complexity

To judge from the traditional methods used to estimate the efficiency of an algorithm, mathematicians are an extraordinarily pessimistic lot. They are the exact opposite of marketers, who try to sell you something based on a best-case estimate of how it will work. Mathematicians, traditionally, have always based their "sales pitch" on a worst-case scenario. If an algorithm runs efficiently 99 times out of 100, but inefficiently one time out of 100, they will tell you it is an inefficient algorithm. Smoothed analysis can be considered to be a very carefully controlled break with this tradition. Put somewhat more technically, the worst-case behavior of the simplex method is exponential, but Spielman and Teng have shown that its "smoothed" complexity is polynomial.

"Exponential" and "polynomial" are the two broadest categories that computational complexity theorists use to classify algorithms. Roughly speaking, exponential algorithms are inefficient and polynomial ones are efficient, although there are many fine gradations within these two categories (and a few gradations outside of them, too, which are irrelevant for linear programming).

An exponential algorithm is one whose run time is an exponential function of the length of the input; for example, given n input bits, it might take 2^n steps to run. Polynomial algorithms have a run time that is a polynomial function of the input length, such as n^{12}. Polynomial algorithms are, in general, much preferred for a very simple reason: Exponentials grow much more quickly than polynomials. Although n^{12} is larger than 2^n for small numbers such as $n = 2$ or 10, the relationship reverses dramatically for larger numbers: 100^{12}, for example, is comparable to the number of molecules in a tablespoon of water, whereas 2^{100} is more like the number of water molecules in a backyard swimming pool. For a number like $n = 1000$ the difference is even more extreme: 1000^{12} is comparable to the amount of water in a large lake, while 2^{1000} is already many orders of magnitude greater than the number of particles in the universe. Thus, generally speaking, a polynomial algorithm scales up well to larger amounts of input data, but an exponential algorithm does not. Little wonder, then, that computer scientists and practitioners look long and hard for polynomial algorithms.

There is one more wrinkle to consider in classifying the complexity of algorithms. Not all kinds of input data are the same. For instance, if you have written a program to sort n numbers in increasing order, your program might run very quickly if it is given n already-sorted numbers as the input. Thus, in practice, different "instances" of a problem may take different amounts of time to solve. There may be an entire landscape of possible input variables and efficiencies, as illustrated in Figure 1. (Usually there are far more than two variables, so this figure is only a schematic representation.) The traditional approach is to consider an algorithm to be no better than its performance at the highest point in the efficiency landscape.

By this criterion, the simplex method is the wrong way to solve a linear programming problem. Until Spielman and Kelner's most recent work, all known versions of the simplex method (there are many of them, as explained below) required exponential time in the worst-case scenario. On the other hand, a different algorithm called the interior point algorithm could be shown to run in polynomial time in all cases. Nevertheless, computer scientists went right on using the simplex method. Were they not listening to the mathematicians? Or did they know something the mathematicians didn't? This was the question that Spielman and Teng set out to answer, back in 1998.

As Simplex As It Gets

To explain Spielman and Teng's result, it helps to back up and explain what linear programming problems are about, and how the simplex method goes about solving them.

Linear programming is a type of constrained optimization problem. (The term "programming" is a remnant of World War II military planning jargon.) The objective in any constrained optimization problem is to maximize or minimize some function of several variables in settings where the variables are required to obey certain inequalities. Such problems are a staple of undergraduate calculus courses. The problems in linear programming, however, are both simpler and more complicated. They are more complicated because they involve many more variables. On the other hand, the objective function and all the inequalities in a linear programming problem are assumed to be linear. They are much simpler than the functions typically encountered in calculus. Indeed, this assumption moves the problem from the realm of calculus to the realm of linear algebra. The generic linear programming problem is to find a vector x that maximizes an inner product $c \cdot x$ subject to a set of inequalities written $Ax \geq b$.

The input of a linear programming problem consists of the components of the vectors b and c and the entries of the matrix A. The crucial parameters are the number of rows and columns of A: Rows count the number of inequalities, and columns count the number of variables. The number of bits to specify all the inputs is another, technical parameter, but for the most basic kind of complexity analysis it is usually ignored.

Each inequality specifies the points on one side of a hyperplane in R^d, where d is the number of variables (also called the dimension of the problem). The intersection of all these sets is a convex, d-dimensional polytope, a generalization of the familiar 2-D polygon and 3-D polyhedron. Linearity of the objective function guarantees that its maximum value occurs at a vertex of the polytope. So in principle linear programming is straightforward to solve: Just find all the vertices of the polytope, compute the objective function at each, and pick the biggest.

The problem with this approach is that the number of vertices can be huge. For example, the inequalities $0 \leq x_i \leq 1$ for $i = 1, 2, \ldots, d$ specify the d-dimensional hypercube, which has

> **On the other hand, a different algorithm called the interior point algorithm could be shown to run in polynomial time in all cases.**

2^d vertices. Right away, one can see that the "look at every vertex" algorithm is doomed to exponential growth. On the other hand, for the hypercube, there is another method for maximizing the objective function $c \cdot x$ that requires essentially no computation: set $x_i = 1$ for the components of c that are positive and $x_i = 0$ for those that are negative. This is an extreme example, but it shows that coming up with a strategy is a good thing. However, for more general polytopes the strategy has to be a good deal more subtle.

The simplex method, introduced in 1947 by George Dantzig, is based on a neat fact about linear objective functions and convex polytopes: If x is a vertex of the polytope and the objective function is not maximized at x, then at least one of the vertices adjacent to x has a strictly larger value of the objective function. Thus one can arrive at the optimal vertex in steps that always increase the objective function.

Dantzig's key contribution was a computational method for stepping from one vertex to the next, using techniques from linear algebra. As mentioned earlier, there are many different versions of the simplex method, corresponding to different strategies for choosing the next vertex. In general, each non-maximal vertex is adjacent to several vertices that improve the objective function. It may seem clear that one should simply move to the one that gives the most improvement, but that's not necessarily the case (see Figure 2). Besides, looking at all adjacent vertices can be a big computation in its own right—it may be just as well to take the *first* vertex that offers any improvement. The next-vertex choice is called a pivot rule; in Dantzig's method, the pivot rule is based on finding the largest entry in the inequality matrix A.

Together with advances in computer hardware, the simplex method revolutionized the science of decision making. Businesses ranging from the airline industry to big-chain retailers began ringing up greater profits—and offering better service—simply by solving equations. The solutions might tell them the optimum way to schedule airline flights, or the optimum way to distribute products to different warehouses. As computers grew more powerful and researchers streamlined the simplex algorithm, practitioners were able to tackle increasingly large problems. It is not unusual these days for a linear programming problem to involve tens of thousands of unknowns and millions of constraints. Indeed, before the Internet became the number-one use of computers, it was estimated that more than half of all computer cycles were devoted to running the simplex method.

How Complex is the Simplex Method?

In 1972, Victor Klee at the University of Washington and George Minty at Indiana University showed that Dantzig's simplex method is worst-case exponential. They did so by giving examples where the algorithm winds up visiting every single vertex in the polytope. Similar examples were found for other pivot rules. In 1996, Günter Ziegler at the Technische Universität in Berlin and Nina Amenta at Xerox PARC in Palo Alto, California, gave a unified construction of "deformed products" of polytopes that produces worst-case examples for all the standard pivot rules.

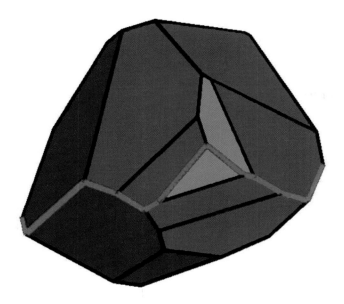

Figure 2. *Schematic diagram of the simplex method. In a linear programming problem, the domain of possible solutions is always a polytope in a high-dimensional space. The problem could be as simple as finding the rightmost point in the polytope. The simplex method works by starting at one vertex of the polytope and hopping to adjacent vertices, always moving to the right (see red path). In general the solution will be found quickly. In some highly unusual cases it will take a long time, and the red path will visit virtually every corner of the polytope. (Figure courtesy of Daniel A. Spielman.)*

The Klee–Minty and other examples made theorists wonder why the simplex method seems to work so well in practice. The first attempts at explaining it were based on so-called average-case complexity analysis. Average-case analysis assumes there is some sort of probability distribution on the set of problems of a given size. The run time of an algorithm is then averaged using the chosen distribution. In 1977, Karl Heinz Borgwardt at the Technische Universität in Kaiserslautern showed that when the simplex method with a pivot rule called the shadow vertex rule is averaged over problems using a Gaussian-type probability distribution, the result is polynomial. In other words, he assumed that the row vectors of the matrix A, as well as the vector c, are independently generated at random in such a way that short vectors are more likely to be generated than long vectors (with the probability of a vector having length r being given by the familiar "bell-shaped curve"). Other researchers, including Stephen Smale at the University of California at Berkeley, found similar results for other pivot rules and probability distributions.

The drawback to these average-case analyses is that they in effect stipulate what a "typical" linear programming problem should be—and their notion of typical bears little relation to what people see in practice. In a sense what the average-case analyses said was that if you take the completely trivial linear programming problem with $A = 0$ and add "noise" to it, then no matter how much noise you've added, the result, on average, takes polynomial time to solve.

Spielman and Teng's smoothed analysis goes a crucial step further. Instead of adding noise to the trivial linear programming problem, they wondered, what if you add it to one of the bad examples? What is the absolute worst that can happen? The answer: even the absolute worst is still polynomial time. The only cost is that the estimate of the running time, which previously depended only on the number of columns and rows of A, now includes a new parameter, namely the variance or "strength" of the noise. Their paper, which appeared in 2004 in the *Journal of the ACM* (Association for Computing Machinery), is 79 pages of combinatorics, probability theory and multi-dimensional integration.

Actually, the route the two mathematicians took to their final result was almost as tortuous as the simplex method's path on a Klee–Minty polytope. Spielman's MIT office is lined with notebooks he kept of the work he and Teng did, starting in 1998. "Many of these notebooks are full of stuff that doesn't work," he says.

"We were originally trying to come up with a simplex method that we could prove took polynomial time," Spielman explains. There came a point, though, "when I realized that what we were doing wasn't going to work."

They had started by studying the constructions of Ziegler and Amenta, trying to visualize their polytopes on the computer. (Teng, who was working at the time for the computer company Akamai, had trained as a computational geometer.) It was "frustrating," Spielman recalls, because the constructions "looked really strange. It was actually very difficult to visualize these polytopes."

Their frustration led to the key insight: "If these bad examples look so strange, maybe that means they really are strange, in the sense that if you assume a little bit of noise in the data, you just don't get them," Spielman says. "Maybe they sort of correspond to configurations that you just don't get under normal circumstances."

It was still an uphill slog. "Once we thought to do it, it was many years before we worked out the proof," Spielman says. "My wife describes it as the smoothed analysis years." (The name "smoothed analysis" was suggested by Spielman's colleague Alan Edelman, an expert on computational linear algebra at MIT.)

Spielman and Teng decided to analyze the shadow vertex pivot rule, which Borgwardt had used in his average-case analysis. The shadow vertex rule projects the d-dimensional polytope down to a 2-D polygon at each step, much as shining a light on a polyhedron casts a polygonal shadow onto a wall or floor. The projection is carried out so that both the current vertex and the eventual maximizing vertex are vertices of the shadow (see Figure 3). The simplex method steps to the vertex on the polytope whose shadow comes next on the polygon's boundary. Because the boundary is one-dimensional, there is no ambiguity about which vertex comes next, and the method is guaranteed to reach the maximizing vertex eventually. To control the number of steps, one has to show that almost all of the vertices of the many-dimensional polytope project down to the middle of the shadow, rather than its edge.

Of course, in the worst case that isn't true. But it turned out that even jostling the worst case a little bit was enough to send most of the projected vertices into the interior of the shadow polygon. As in Borgwardt's analysis, the "jostling" corresponds to adding a small random vector to each row of A, but this time starting with a worst-case A instead of the trivial case $A = 0$. Spielman and Teng then averaged the result over all of the possible random vectors (whose length cannot get very big because it is controlled by a bell-shaped probability distribution). They found themselves staring at some nasty-looking multi-dimensional integrals.

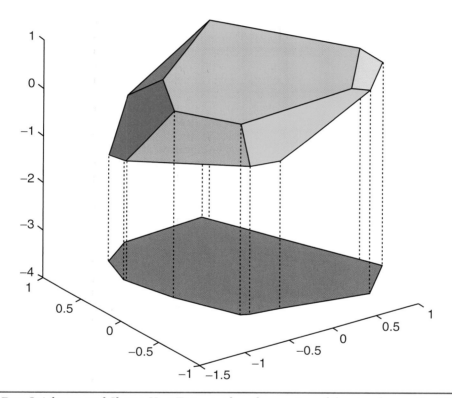

Figure 3. *Dan Spielman and Shang-Hua Teng analyzed a version of the simplex method that projects the high-dimensional polytope down to its "shadow" in a 2-dimensional plane. This eliminates ambiguity about which adjacent vertex to move to next, because all of the vertices have a cyclic order. Spielman and Teng showed that this version of the simplex method has a smoothed polynomial-time complexity. (Figure courtesy of Daniel A. Spielman.)*

At this point, an epiphany occurred. Ten years earlier, one of Spielman's professors, Gian-Carlo Rota, had given him a book about geometric probability. "I don't know what inspired him to give that to me," Spielman says. He had paid little attention to it at the time. But after a long period of fruitless struggle, he realized that a formula for the exact type of integral he was trying to do was proven in the book—it was called Blaschke's formula. Though it sounds abstruse, at its heart it involves nothing

As long as computer programmers have to work with noisy data, smoothed analysis should be a powerful way to make the inherent randomness of the data work in their favor.

more complicated than the change-of-variables formula from several-variable calculus. "I'm embarrassed to say how long it took me to realize that," Spielman says.

The basic proof was complete in 2001. Since then, Spielman and his students have continued to refine their methods. "We have become much more sophisticated about the changes of variables we use," Spielman says. Not surprisingly, the first one he and Teng stumbled into by accident was not the best. Finally, in May 2006, he and Kelner were able to announce the first version of the simplex method that runs in polynomial time, even in the worst case. Actually, there is some debate over whether it literally qualifies as a simplex method. Spielman first transformed the linear program to a related problem whose solution is not affected by small random perturbations. They then added a little bit of noise to the input data. That was enough to guarantee that the solution would be found in a reasonable amount of time, without changing the combinatorial structure of the problem. For purists, this method does not count as a simplex algorithm because it does not work with the original polytope. Nevertheless, its speed makes it an attractive method of solution.

Meanwhile, smoothed analysis continues to make rapid progress, with applications to other kinds of algorithms. For example, Sergei Vassilvitskii and David Arthur, theoretical computer scientists at Stanford University, have derived smoothed-complexity estimates for a well-known method used in image compression and data analysis. The k-means method is a way of separating data points in a high-dimensional space into k "clusters" of nearby points. For example, the 32-means method could be used to compress an image with 256 colors down to 32 colors. The algorithm would find the optimal way to cluster the pixels in the image into 32 color groups, minimizing the (squared) discrepancies between the selected colors and the originals.

As with the simplex method, practitioners have found that the k-means method works far better than it is supposed to. Vassilvitskii and Arthur have now proved that the smoothed complexity is independent of the dimension of the space in which the data points lie, and polynomial in the number n of data points. In fact, the smoothed complexity is on the order of n^k. They are hoping to do even better because in practice the time does not even seem to depend on k; however, that is still a goal for future research.

As long as computer programmers have to work with noisy data, smoothed analysis should be a powerful way to make the inherent randomness of the data work in their favor. "It's a very natural idea," says Vassilvitskii. "It's not as if you were playing a game with the devil, who is feeding you the worst possible numbers."

Domino Theory

Linear programming is just one of many constrained optimization problems with important applications. Among the others is one called *integer* programming. The formulation, in terms of linear inequalities and a linear objective function, is the same as for linear programming. The key difference is that the sought-for optimum is required to have integer components.

Integer programming arises in innumerable business applications, ranging from airline crew scheduling to capital investment budgeting. (Suppose, for example, you have a certain budget for opening new stores in various locations. The cost of opening each store varies, but so does the projected revenue stream. This can be posed as the problem of maximizing total revenue $r_1x_1 + r_2x_2 + \cdots + r_nx_n$ subject to $c_1x_1 + c_2x_2 + \cdots + c_nx_n \leq B$, but with the additional stipulation that each x_i either be 1 or 0—that is, you either open a store at location i or you don't. Similarly, in crew scheduling, each pilot either is or isn't assigned to a particular flight.)

Unlike linear programming, there are no known polynomial time algorithms for integer programming. In fact, integer programming belongs to the class of intrinsically hard problems known as NP-complete problems, for which the existence of a polynomial time algorithm would automatically translate into polynomial time algorithms for a vast array of other problems (see ∞). Nevertheless, computer scientists have found practical methods that enable them to solve surprisingly large integer programming problems.

Robert Bosch at Oberlin College in Oberlin, Ohio, has found a novel, artistic application of integer programming. He uses it to design portraits made of sets of dominoes (see Figure 4, next page). The idea of domino portraits is to take several sets of dominoes and, using *all* the dominoes, arrange them so as to suggest the gray- scale shading of a black-and-white picture. Bosch was inspired by the work of Ken Knowlton, a computer graphics pioneer at Bell Labs, who did the first domino portraits in the 1980s.

In Bosch's integer-programming approach, there is a variable for each possible placement of each kind of domino on a $10m \times 11n$ grid, using mn complete sets of dominoes. (Each complete set consists of 55 dominoes, corresponding to number pairs (a, b) with $0 \leq a \leq b \leq 9$.) There are two types of constraint. One stipulates that each type of domino is used exactly mn times; the other stipulates that each grid cell is covered by exactly one domino. The "cost" of each variable is determined by the extent to which placing the corresponding domino in the corresponding location fails to match the gray scale of the picture. For example, placing the blank domino in a white portion of the picture is more expensive than putting the double-9 domino there. (In some of his work, Bosch uses "reverse" dominoes, with black pips on white rectangles, in which case costs are reversed.)

Bosch has done portraits with as many as 100 sets of dominoes ($m = n = 10$). The integer programming problem in such cases has 2,179,000 variables and 11,055 constraints. Nonetheless, the optimal solution is obtainable on a desktop computer.

The idea of domino portraits is to take several sets of dominoes and, using *all* the dominoes, arrange them so as to suggest the gray-scale shading of a black-and-white picture.

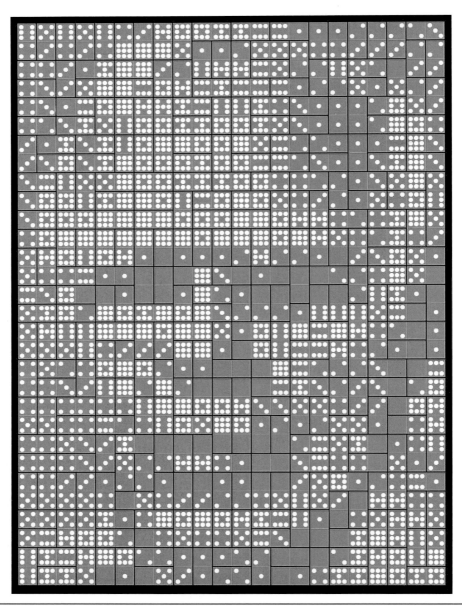

Figure 4. *A "domino portrait" of Martin Gardner, the famed mathematical writer. The creation of such a portrait can be reduced to a problem in integer programming. ("Martin Gardner, 6 sets of 9-9 dominoes,"* © *1993 Ken Knowlton.)*

WHAT'S HAPPENING IN THE MATHEMATICAL SCIENCES